Third Edition

Number Treasury³

Investigations, Facts and Conjectures
about More than 100 Number Families

Third Edition

Number Treasury³

Investigations, Facts and Conjectures about More than 100 Number Families

Margaret J Kenney • Stanley J Bezuszka

Boston College, Massachusetts, USA

 **World Scientific**

NEW JERSEY • LONDON • SINGAPORE • BEIJING • SHANGHAI • HONG KONG • TAIPEI • CHENNAI

Published by

World Scientific Publishing Co. Pte. Ltd.

5 Toh Tuck Link, Singapore 596224

USA office: 27 Warren Street, Suite 401-402, Hackensack, NJ 07601

UK office: 57 Shelton Street, Covent Garden, London WC2H 9HE

Library of Congress Cataloging-in-Publication Data

Kenney, Margaret J.

  Number treasury 3 : investigations, facts, and conjectures about more than 100 number families / by Margaret J. Kenney (Boston College, USA) and Stanley J. Bezuszka (Boston College, USA). -- 3rd edition.

    pages cm

  Includes bibliographical references and index.

  ISBN 978-9814603683 (hardcover : alk. paper) -- ISBN 978-9814603690 (softcover : alk. paper)

  1. Numeration. 2. Mathematical recreations. I. Bezuszka, Stanley J., 1914–2008. II. Title. III. Title: Number treasury three.

  QA141.K46 2015

  513.5--dc23

                                                                2014040942

British Library Cataloguing-in-Publication Data

A catalogue record for this book is available from the British Library.

Typeset by Stallion Press

Email: enquiries@stallionpress.com

Printed in Singapore

Number Treasury[3] is dedicated to my co-author Stanley J. Bezuszka, longtime Professor of Mathematics and Director of the Mathematics Institute at Boston College, who was a source of inspiration and encouragement to so many in the world of mathematics education.

Margaret J. Kenney

Contents

Foreword

The gift of numbers like the gift of fire has made the world much brighter.
                                                    Stanley J. Bezuszka (1914–2008)

Introduction

Time and numbers were born together. Time is a measure of change, and numbers express that measure. There is an awesome mystery that surrounds the all-pervasive dimension of time. Likewise, there is an awesome mystery and fascination about numbers that attract distinguished mathematics researchers as well as imaginative amateurs the world over. Teachers who pursue numbers and the rich history of numbers with their classes can provide students with an understanding that mathematics is a collaborative effort that has been nurtured by individuals and groups representing many cultures and periods of history. Students can learn to become accomplished investigators, to make discoveries, and contribute to a branch of mathematics that is vibrant and motivating. **Number Treasury**[3] has evolved in order to serve as a catalyst for those who ascribe to this point of view.

Details

Number Treasury[3] is a broadening and update of **Number Treasury**[2]. The book contains information about more than 100 families of positive integers. Brief historical notes often accompany the descriptions and examples of the number families. Exercises for each major family are provided to stimulate insight. Some exercises contain problems that are thought provokers to be resolved simply with paper and pencil; others should be tackled with calculator in hand so that lengthier computations can be managed with ease and take the results to a higher level of understanding. Still other problems are intended for more extensive exploration with the use of computer software. In some instances it is helpful to model problems with hands-on materials.

The emphasis in **Number Treasury**³ is on doing rather than proving. However, the reader is urged to think critically about situations, to provide reasoned explanations, to make generalizations and to formulate conjectures. The book begins with a chapter of Investigations. These are principally stand-alone activities that represent content drawn from the Chapters 2 through 7 of the book. Their purpose is to set the tone of the book and to stimulate student reflection and research in a variety of areas. In fact, throughout the book, the reader will find numerous open-ended problems. This book also contains detailed solutions to the Exercises and Investigations. A Glossary and Index are provided for quick access to information. References and recommended readings are supplied so that teachers and students can use this book as a stepping stone to more concentrated study.

Who Uses Number Treasury³

This book is written for teachers and students. For teachers **Number Treasury**³ is a resource for instructional preparation and problems, together with snapshots of mathematical history intended for teachable moments. For students who are engaged in learning about number families and who are assigned problems, projects and papers, **Number Treasury**³ is a useful source of ideas and topics. The mix of discussion with examples and illustrations is intended to serve as a writing model for the student. Both audiences should think critically about the content, provide carefully reasoned explanations, make generalizations, and form conjectures.

Who Is Involved

The first edition was completed with the able assistance of six Boston College graduates and undergraduates: Jeanne Cavanaugh, James Cavanaugh, Claudia Katze, Stephen Kokoska, Jill Nille, and Jonathan Smith.

Seven Boston College graduates and undergraduates were indispensable in the production of the second edition. Special thanks and grateful appreciation go to Joan Martin for her thoughtful content and style suggestions, editorial advice and word processing skills; to Cynthia Tahlmore, Geraldine Mele, and Erin Mitchell for computer graphics and word processing assistance; to Allyson Russo, Shannon Toomey, and Megan Mazzara for problem solutions.

The third edition has been completed by the surviving original author with the invaluable assistance and perseverance of Geraldine Mele who offered not only content suggestions but who also especially contributed word processing, computer graphics and style expertise. Sincere gratitude and appreciation is also extended to Joan Martin for her careful review of the manuscript.

Chapter 1

A Perfect Number of Investigations
$28 = 1 + 2 + 4 + 7 + 14$

A GREAT discovery solves a great problem but there is a grain of discovery in any problem.
George Polya (1887–1985)

What are They?

The Investigations that follow are a set of stand-alone activities. Each Investigation focuses on at least one number family or topic relating to numbers. All but six of the Investigations are described on one page. Share with students that the problems in an Investigation are intended to be challenging in many ways:

- The time needed to complete an Investigation may vary and exceed the time required to finish a typical homework assignment.
- The computation necessary to bring closure may be lengthy and demanding — even with the use of technology.
- The amount of writing, discussing, explaining and illustrating may be more than anticipated.

Teacher Tips

The Investigations are listed in ascending order of difficulty in the table on the next two pages. There are three levels of difficulty represented in the 28 Investigations that can be assigned individually or adapted for group work. The lowest level consists of the first eight Investigations that have the least prerequisites. The middle level consists of the next nine Investigations and requires more use of abstract reasoning and familiarity with algebraic expressions. The final 11 Investigations challenge the student to persist and probe more deeply in order to complete the work.

There is also a column in the table naming the most significant prerequisite(s) needed by the student to understand and carry out the work in each Investigation. The teacher may choose to provide additional content background for some specific Investigations prior to assigning them. Assign the Investigations as extended homework or as in-class work. Some Investigations call for the preparation of reports. Thus, students may need further directions, especially about the kind of resources available for them to use. Students should know the Internet is an excellent resource, and that it should be used appropriately as they compile their reports.

Finally, pages noted in the Prerequisites column refer to related material contained in Chapters 2 through 7.

Footsteps of Lagrange

Develop the first 20 terms of the sequence s_n, where s_n is the number of ways n can be written as a sum of at most 4 squares. Note one term will be counted as a sum.

EXAMPLE

How many ways can 5 be written as a sum of at most 4 squares?
1 way since $5 = 4 + 1 = 2^2 + 1^2$.

EXAMPLE

How many ways can 9 be written as a sum of at most 4 squares?
2 ways since $9 = 2^2 + 2^2 + 1^2$ and $9 = 3^2$.

1. Fill in the following table.

n	Ways to Write n as Sum of at Most 4 Squares	Number of Ways
1		
2		
3		
4		
5	$2^2 + 1^2$	1
6		
7		
8		
9	$3^2; 2^2 + 2^2 + 1^2$	2
10		
11		
12		
13		
14		
15		
16		
17		
18		
19		
20		

2. Describe in a few sentences some patterns you observe in the table.
3. Find a number and verify that it can be written as a sum of at most 4 squares in exactly
   a) three ways   b) four ways

Trying Trapezoids

Trapezoidal numbers are figurate numbers that can be represented by trapezoidal arrays of two or more rows of dots or chips.

For example, the number 5 can be pictured as a trapezoidal number using two rows of dots or chips:

Isosceles array   Right-angled array

There are many different ways to describe trapezoidal numbers. Two of these are shown below.

- Subtracting triangular numbers produces trapezoidal numbers.

EXAMPLE

Start with T_4. Subtract T_1 and T_2.

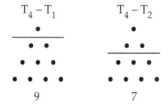

Thus 7 and 9 are trapezoidal numbers.

1. Using subtraction of triangular numbers procedure, name the trapezoidal numbers that come from

    a) T_5 b) T_8 c) $T_k, k > 2$

- Adding groups of consecutive numbers greater than 1 produces trapezoidal numbers.

EXAMPLE

Trapezoidal numbers that consist of 3 rows, one of which has 4 dots, are

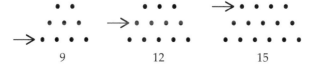

2. Name the trapezoidal numbers that consist of
   a) 4 rows one of which has 3 dots.
   b) 5 rows one of which has 4 dots.
   c) 3 rows one of which has k dots, $k > 1$.
3. List all ways that each given number can be trapezoidal.

   a) 11   b) 27   c) 33

4. The trapezoidal number 9 can be pictured in isosceles or right-angled form:

         isosceles   right-angled

   Study the right-angled model and develop a formula for a trapezoidal number in terms of the number of rows and the first and last rows.

5. When expressed in right-angled form it is clear that a trapezoidal number in the sequence $2 + 3, 2 + 3 + 4, 2 + 3 + 4 + 5, \ldots$ can be written as the sum of a $2 \times n$ rectangular number and a triangular number.
   For example,

       $2 + 3 + 4 = 6 + 3$

   Write each number as the sum of a $2 \times n$ rectangular number plus a triangular number.

   a) $2 + 3 + 4 + 5$ c) $2 + 3 + 4 + \cdots + 10$
   b) $2 + 3 + 4 + 5 + 6$ d) $2 + 3 + 4 + \cdots + 20$

6. Segments joining the dots in the trapezoidal number $2 + 3 + 4 = 9$ form an array
   1 hexagon with 2 dots per side and 2 triangles.

   Determine that segments joining the arrays of dots

   a) $2 + 3 + 4 + 5 + 6$ forms 3 hexagons and 6 triangles
   b) $2 + 3 + 4 + 5 + 6 + 7 + 8$ forms 6 hexagons and 12 triangles.

   Predict, then verify, the number of hexagons and triangles for the arrays of dots

   c) $2 + 3 + \cdots + 10$
   d) $2 + 3 + \cdots + 12$
   e) $2 + 3 + \cdots + 20$

Hexagons in Black and White

1. To produce the patterns below you will need a quantity of two different colored chips. Copy each of the designs shown and then create the next two that follow the pattern. Use the designs to produce the data in the table. Let n be the number of hexagon paths around the center chip.

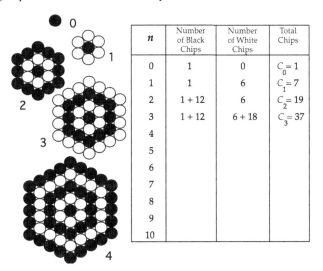

n	Number of Black Chips	Number of White Chips	Total Chips
0	1	0	$C_0 = 1$
1	1	6	$C_1 = 7$
2	1 + 12	6	$C_2 = 19$
3	1 + 12	6 + 18	$C_3 = 37$
4			
5			
6			
7			
8			
9			
10			

A *recursive* formula for a sequence defines each term of the sequence using the preceding term or terms.

An *explicit* formula for a sequence defines each term of the sequence by a rule that depends only on the term number.

2. a) Write a recursive relation for the total number of chips C_n in row n of the table. That is, express C_n in terms of C_{n-1}.

   b) Give an explict formula $f(n)$ for the total number of chips in terms of n.

3. Create your own patterns using two different colored chips. Gather the data and record it. Summarize your findings with a recursive and an explicit formula.

Marble Art

Polyhedral numbers can be represented physically in a variety of ways. For example, you can produce effective and attractive models from practice golf balls, holiday glass ball ornaments, or marbles held together with glue. Two models using marbles are described below. Try these and then create some variations of your own.

Square pyramidal numbers

1. Make glued layers of marbles representing the consecutive square numbers. Let them stand until they are completely dry.

   1×1 2×2 3×3 4×4 5×5

2. Start with the 5×5 layer. Place and glue the 4×4 layer on top so that it rests in the indentations of the 5×5 layer. Continue by placing and gluing the 3×3 layer, the 2×2 layer, and so on. When all layers are glued in place, let the model stand until dry.

3. For a better effect you can glue 4 single marbles at each of the corners to serve as feet so that the model is raised from the surface. A mirror can be positioned under the model to provide interesting reflections.

Tetrahedral cluster

1. Build and glue a hexagonal array that has 3 marbles per side. Let it stand until completely dry.

2. Create 6 feet for the hexagonal array, 3 feet each for the top and bottom sides of the surface. Each foot here is composed of 4 marbles (1 marble glued onto the indentation of a triangular array of 3 marbles).

Foot

3. Apply glue and attach the three feet to each side of the hexagonal layer as shown in the figure. Attach one additional marble where the 3 feet meet in the center on each side of the hexagonal array as shown in the diagram. Let the model stand until completely dry.

 Top and bottom view

Honest Number Hunt

> An *honest number* in a language is a number whose word letter count and size are equal.

In English, "four" represents as many objects as there are letters in the word four. In fact, four is the only honest number in English. In Spanish, "cinco" represents as many objects as there are letters in the word cinco. Cinco is an honest number in Spanish.

1. Undertake an extensive search for honest numbers in as many languages as possible. Prepare a chart with your class that displays the honest numbers found. Do all the languages you examined have at least one honest number?

2. Honest numbers in a language have an interesting property as the following example illustrates.

EXAMPLE   English

Choose a number.	2020
Write the number word for 2020.	two thousand twenty
Count and record the number of letters.	17
Write the number word for 17.	seventeen
Count and record the number of letters.	9
Write the number word for 9.	nine
Count and record the number of letters.	4
Write the number word for 4.	four

If you continue, **4** and **four** keep repeating.
The number trail has 4 numbers and ends in the honest number 4.

$$2020 \longrightarrow 17 \longrightarrow 9 \longrightarrow 4$$

a) Using the English language, apply this process to other numbers. Summarize your findings.

b) Choose another language in which you know how to count and which has an honest number. Apply the process described in the example to several numbers in this language. In a paragraph or two, summarize your findings for a class report.

Seeking Honesty in Numbers

> An *honest number* in a language is a number whose word letter count and size are equal.

Ten represents 10 objects, but 10 has 3 letters. Ten is not an honest number.
Four represents 4 objects and four has 4 letters. Four is an honest number.
In the English language, 4 is the only honest number.

Determine which number trails end in the honest number 4.

EXAMPLE

Choose a number.	1066
Write the number word for 1066.	one thousand sixty-six
Count and record the number of letters.	19
Write the number word for 19.	nineteen
Count and record the number of letters.	8
Write the number word for 8.	eight
Count and record the number of letters.	5
Write the number word for 5.	five
Count and record the number of letters.	4

If the steps are continued, **4** and **four** keep repeating.
This number trail has five numbers and ends in 4.

$$1066 \longrightarrow 19 \longrightarrow 8 \longrightarrow 5 \longrightarrow 4$$

Find the number trails for:

1. 63
2. 163
3. 999
4. 8999
5. any five-digit number
6. any six-digit number
7. any seven-digit number
8. any eight-digit number

With your class, produce a table or diagram that shows all possible number trails for the first 200 integers.

Geoboard Journeys

Use a geoboard of size 5 pegs by 5 pegs or larger, together with a sheet of square dot paper for recording results.

Your task: Count the number of paths from A to B that are composed of only horizontal or vertical segments. You may move only to the right or up. Remember to use only pegs that are on or below the diagonal from A to B.

EXAMPLES

On a 2×2 geoboard there is 1 path.

On a 3×3 geoboard there are 2 paths.

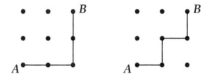

In problems 1 and 2, A is the lower left peg and B is the upper right peg.

1. Produce the acceptable paths from A to B on a
   a) 4×4 geoboard   b) 5×5 geoboard
   and record the results on dot paper. Try to be systematic in recording solutions.
2. Based on the solutions for the 2×2, 3×3, 4×4, and 5×5 sizes, identify if you are able the kind of numbers that represent the total path count from A to B.
3. Using only horizontal or vertical segments and any pegs for B, count all acceptable paths from A to B. Give the total and describe your counting strategy on a
   a) 2×2 geoboard   b) 3×3 geoboard
   c) 4×4 geoboard   d) 5×5 geoboard

4. Can you discover a connection between your solutions to problems 2 and 3 and the numbers in Pascal's triangle? Explain your thinking.
5. Create and solve your own geoboard journey problem.

Mysterious Mountains and Binary Trees

Mountains are constructed with up-segments and down-segments.

1 mountain design is possible using one up– one down-segment.

2 mountain designs are possible using two up–two down-segments.

1. Draw all mountain designs composed of 3 up-segments and 3 down-segments.
2. Draw all mountain designs composed of 4 up-segments and 4 down-segments.

A rooted binary tree starts from a vertical edge. Rooted binary trees will be ordered by the number of *interior* vertices. One binary tree is different from another if it cannot be turned to match the other.
Here are some rooted binary trees.

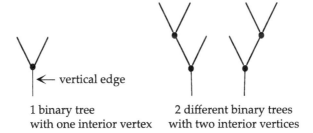

← vertical edge

1 binary tree with one interior vertex

2 different binary trees with two interior vertices

3. Draw all different rooted binary trees with 3 interior vertices.
4. Draw all different rooted binary trees with 4 interior vertices.
5. Describe any connection you can make between the mountains and the trees.

Fermat Factorings

Pierre de Fermat (1601–1665) a French lawyer by profession and an amateur mathematician by choice developed an algorithm for finding the prime factorization of an integer. Fermat's algorithm is applied to odd integers that are not squares. His strategy consisted in adding consecutive odd integers to N until a square was reached. If the given number is even, first divide by 2 until an odd non-square integer N appears.

EXAMPLE

Find the prime factorization of 105.

$$
\begin{array}{r}
105 \\
+\,1 \\
\hline
106 \\
+\,3 \\
\hline
109 \\
+\,5 \\
\hline
114 \\
+\,7 \\
\hline
121 = 11^2
\end{array}
$$

Since 121 is a square, then 105 can be written as $(11 + \_)(11 - \_)$, where $\_$ is replaced by the *number* of odd integers needed to reach a perfect square. In this case, 4 odd integers are added. Thus $105 = (11 + 4)(11 - 4) = 15 \times 7$. Now factor 15 by this procedure.

$$
\begin{array}{r}
15 \\
+\,1 \\
\hline
16 = 4^2
\end{array}
$$

So $15 = (4 + \_)(4 - \_) = (4 + 1)(4 - 1) = 5 \times 3$

Finally, $105 = 3 \times 5 \times 7$.

EXAMPLE

Find the prime factorization of 84.

Since 84 is even, divide by 2 until an odd non-square integer appears.

$$84 = 2 \times 42 = 2 \times 2 \times 21$$

$$
\begin{array}{r}
21 \\
+\,1 \\
\hline
22 \\
+\,3 \\
\hline
25 = 5^2
\end{array}
$$

Then $21 = (5 + \_)(5 - \_)$
$= (5 + 2)(5 - 2)$
$= 7 \times 3$

Finally,
$84 = 2^2 \times 3 \times 7.$

1. Use Fermat's algorithm to find the prime factorization of

   a) 185   b) 360   c) 237   d) 500

2. Explain why Fermat's algorithm works.
3. Write a report on the life of Pierre de Fermat, including names of his mathematical friends and some of the mathematical problems he worked on. Be sure to include information about Fermat primes, Fermat's Last Theorem, and Andrew Wiles.

Factor Lattices

The collection of divisors of an integer can be represented using vertices and line segments, in linear, two, three, and higher dimensional arrays. Such figures are called factor lattices and each vertex of these figures is labeled with one of the divisors of the integer. Factor lattices can be produced using prime factorization and the concept of least common multiple.

The factor lattice for any number that is a power of a prime is linear. For example, the factor lattice for $81 = 3^4$ is:

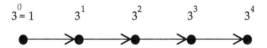

The factor lattice for 81 has 5 vertices, one for each divisor of 81. Also, the factor lattice has four unit segments.

Observe that powers of the prime 3 are matched in order starting with 3^0. Check to see that the factor lattice for 2^7 has eight vertices and seven unit segments.

1. The factor lattices for 75 and 36 are shown below. Note that *lcm* is an abbreviation for *least common multiple*. Study them carefully for clues.

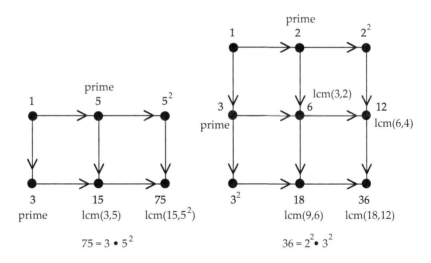

Use them as guides and draw factor lattices for:

a) 24   b) 72   c) 100

Numbers whose prime factorization is a *single* prime to some power have a factor lattice that is *one* dimensional, while numbers whose factorization consists of *two* distinct primes have factor lattices that are *two* dimensional. It appears that a number that has *three* distinct prime factors should have a *three* dimensional factor lattice. Indeed, the factor lattice for 30 can be represented by the vertices and edges of a cube.

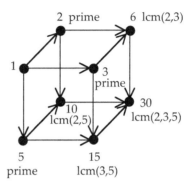

2. Name 5 other numbers whose factor lattice is also a cube. Using the example for 30, draw the cubes and match divisors with vertices.

Try *building* models of three dimensional factor lattices. You can use toothpicks as edges with gum drops or mini-marshmallows for vertices or commercially available materials.

3. Name a number whose factor lattice is the figure shown below.

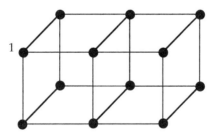

   Check your guess by assigning divisors to vertices.
4. Draw and label the factor lattices for 90 and 168.
5. Match the vertices of a four dimensional hypercube with the divisors of the number 210.

A Juggling Act

You have two unmarked containers that you take to a well.

One container holds a liters, the other b liters of water. Your job is to obtain exactly <u>1</u> liter of water in one of the containers. Explain in detail how you will accomplish this if:

1. $a = 5$ $b = 7$

2. $a = 4$ $b = 7$

3. $a = 5$ $b = 8$

4. Describe in detail how you could use a 5-liter container and an 8-liter container to get each of the amounts $1, 2, 3, \ldots, 13$ liters.

5. If you have a 6-liter and an 8-liter container, can you obtain exactly 1 liter of water? Explain your answer.

6. When a and b are given, how can you tell whether it is possible to obtain exactly 1 liter of water?

7. You have three containers whose capacities are 8, 5, and 3 liters.
   The 8-liter container is full of water. Describe how to split this water you have into two equal amounts.

8 liters   5 liters   3 liters

The Super Sum: $1 + 2 + 3 + \cdots + n$

As a young boy, Carl Friedrich Gauss (1777–1855) attended St. Katherine's Volksschule, a one-room school in Brunswick, Germany. The story is told that when he was about eight years old, his teacher asked the class to compute the sum $1 + 2 + 3 + \cdots + 100$. Carl was able to get the solution quickly and was the first to place his slate with the correct answer on the table.

1. Solve Carl's problem. Describe in detail how you arrived at your total.
2. Use your reasoning in problem 1 to compute

   a) $1 + 2 + 3 + \cdots + 99$ b) $1 + 2 + 3 + \cdots + 101$

   c) $1 + 2 + 3 + \cdots + 500$ d) $1 + 2 + 3 + \cdots + 175$

3. Generalize your reasoning in problem 2 to find the total of the super sum $1 + 2 + 3 + \cdots + n$ when

   a) n is even   b) n is odd

4. Give one expression for the total of $1 + 2 + 3 + \cdots + n$, where n is an odd or even number.
5. Use your answer to problem 4 to compute

   a) $2 + 4 + 6 + \cdots + 200$ b) $3 + 6 + 9 + \cdots + 255$

You can count all rectangles in this 2×3 grid by listing the various size rectangles with their count in a size table. The total number of rectangles is the sum of the numbers in the table. Study the example carefully.

6. Find the rectangle count for the grids of size

   a) 3×7 b) 4×8

7. **Challenge** Use the super sum total to find the total rectangle count for grids of size

   a) $5 \times n$ b) $m \times n$

Conjecturing with Pascal

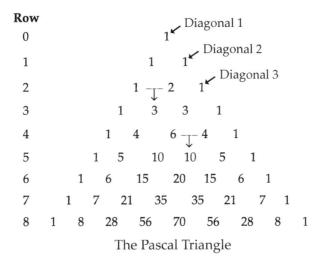

The Pascal Triangle

A *conjecture* is a statement that appears to be true based on the evidence at hand. In the Pascal triangle investigations below, you will get an opportunity to formulate your own conjectures.

1. Calculate the row sums for the rows 0 through 8. Look for patterns to complete the conjecture: In row n, the row sum is ___. Calculate the cumulative row sums for rows 0 through 8. Complete the conjecture: The cumulative row sum for rows 0 through n is ___.

2. a) Alternate $+/-$ between the numbers in each row 0 through 8. Then calculate the sums. For example, $1 + 4 - 6 + 4 - 1 = 2$. Look for patterns to complete the conjecture: In row n, the sum is _____.

   b) Alternate $-/+$ between the numbers in each row 0 through 8. Then calculate the sums. For example, $1 - 4 + 6 - 4 + 1 = 0$. Look for patterns to complete the conjecture: In row n, the sum is _____.

3. a) In row 0 the single number is odd; in rows 1 and 3 each number is odd. Name the next 5 rows in which each number is odd.

   b) In rows 2 and 4, except for the two 1s, the remaining numbers are even. Name the next 5 rows in which each number, except for the two 1s, is even. Explain why this happens.

   c) Based on your findings make a conjecture that indicates which rows contain all odd numbers and which rows except for the two 1s contain all even numbers.

4. Study the hexagonal grid on which the first few rows of the Pascal triangle appear. Without first filling in the numbers in more rows of the triangle, shade the hexagons in each row as follows. If a hexagon represents an odd number shade it, and if a hexagon represents an even number do not shade it. Use these facts:

   a) The sum of two even numbers is an even number.
   b) The sum of two odd numbers is an even number.
   c) The sum of an even and an odd number is an odd number.

   A triangular design should result. By counting unshaded hexagons, identify the various triangular numbers that appear in your design.

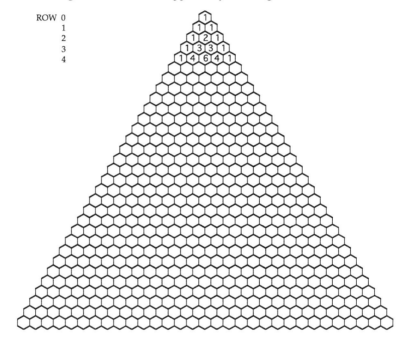

5. Find references that have information on the Sierpinski triangle. Prepare a report that describes how this triangle is formed and how it is related to your Pascal triangle patterns. Find out about the Chaos Game. How does it work? How is it connected to these designs?

Pythagorean Triple Pursuits

A Pythagorean triple is a collection of three positive integers a, b, c such that $a^2 + b^2 = c^2$.

1. a) Use a spreadsheet to list 50 different values for a, b, and c where the n's are positive integers and

$$a = 2n + 1, \qquad b = 2n^2 + 2n, \qquad c = 2n^2 + 2n + 1.$$

   b) Show that $a^2 + b^2 = c^2$ for the values of a, b, and c in 1(a).

2. a) Use a spreadsheet to produce three lists of 50 different values for a, b, and c where m, n are positive integers and $m > n$ and

$$a = 2mn, \qquad b = m^2 - n^2, \qquad c = m^2 + n^2.$$

   Start with $n = 1$ and let $m = n + 1$; $m = n + 2$; and $m = n + 3$. Choose other combinations for m, n and use the spreadsheet to list other sets of 50 different values.

   b) Show that $a^2 + b^2 = c^2$ for the values of a, b, and c in 2(a).

3. Consult numerous paper and Online resources that treat mathematics content as well as the history of mathematics.

   a) Write a brief report about Pythagorean triples that includes an explanation of the difference between a Pythagorean triple and a *primitive* Pythagorean triple.

   b) In your report include several formulas that produce *primitive* Pythagorean triples. Find out when and where the formulas originated.

4. There are 50 Pythagorean triples in which each number is less than 100. Make a list of the ones you find. Discuss the patterns you observe in and among the triples.

Pentagonal Play

1. The first five pentagonal numbers are:

$$P_1 = 1, \quad P_2 = 5, \quad P_3 = 12, \quad P_4 = 22, \quad P_5 = 35.$$

   a) Name the next five pentagonal numbers P_6 through P_{10}.
   b) Find a formula that gives the nth pentagonal number P_n.

2. The pentagonal numbers greater than 1 can be represented in a "houselike form" as the sum of dot arrays that are a combination of a square plus a triangle array.

Pentagonal numbers
Nonregular pentagon arrays

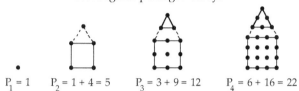

$P_1 = 1 \qquad P_2 = 1 + 4 = 5 \qquad P_3 = 3 + 9 = 12 \qquad P_4 = 6 + 16 = 22$

   Sketch the "houselike forms" for P_5, P_6, P_7. Express P_8, P_9, P_{10} as the sum of a square and triangular number. Express P_n as the sum of a particular square and triangular number.

3. The triangular numbers are $T_1 = 1$, $T_2 = 3$, $T_3 = 6, \ldots$ with $T_n = n(n+1)/2$. Observe that $P_2 = 5 = 3 + 1 + 1$ and $P_3 = 12 = 6 + 3 + 3$. Write each of the pentagonal numbers P_4 through P_{10} as the sum of three triangular numbers. Show that in general P_n can be written as the sum of three triangular numbers.

4. Compute the sum of the digits of each pentagonal number through P_{20}. If necessary, keep adding to reduce the sum to a single digit. This digit is called the digital sum. Which digits occur as digital sums? Make a conjecture about the patterns you observe.

5. List the units or ones digit of each of the first 25 pentagonal numbers. Which digits appear in the list? Make a conjecture about the patterns you observe.

Triangular Number Turnarounds

> A *triangular number* is a number that is the sum of consecutive numbers starting with 1.
>
> $$\underline{1}, \quad 1 + 2 = \underline{3}, \quad 1 + 2 + 3 = \underline{6}, \quad 1 + 2 + 3 + 4 = \underline{10}$$
>
> are triangular numbers. A formula for the nth triangular number T_n is given by $T_n = n(n + 1)/2$.

1. Find the first 3 triangular numbers greater than 100. Explain how you get your answer.
2. Are there any triangular numbers between 200 and 225? Explain how you get your answer.
3. Triangular numbers can be represented by triangular arrays of dots or chips.

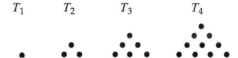

$$T_1 \qquad T_2 \qquad T_3 \qquad T_4$$

Determine the minimum number of dot or chip *moves* required to turn the triangle from a *point up* to a *point down* position.

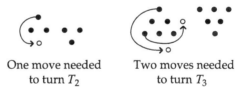

One move needed   Two moves needed
to turn T_2 to turn T_3

Develop a strategy and draw the sketches to turn around $T_4, T_5, \ldots, T_{10}$.
4. Make a spreadsheet or calculator listing of the first 50 triangular numbers. Use this data to estimate if there are just as many even triangular numbers as odd triangular numbers amongst the first 100 triangular numbers. Extend your spreadsheet to verify your guess.

Centered Triangular Numbers

The number below each figure is its dot count.

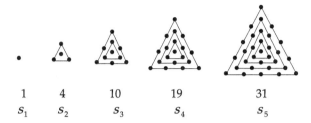

1	4	10	19	31
s_1	s_2	s_3	s_4	s_5

1. Draw the next figure in the sequence of *centered triangular numbers*, the work of Jordanus de Nemore from Germany in the 13th century, and give its dot count.
2. Complete the table below by naming the next four elements of the sequence s_n.

n	1	2	3	4	5	6	7	8	9	10
Dot Count s_n	1	4	10	19	31					136

3. Find a recursive formula for s_n, the number of dots in the nth centered triangular number.
4. Find an explicit formula for s_n, the number of dots in the nth centered triangular number.
5. Draw and give the dot count for the first five figures in the sequence q_n, of centered squares.
6. Complete the table below by naming the elements of the sequence q_n.

n	1	2	3	4	5	6	7	8	9	10
Dot Count q_n	1	5								181

7. Find a recursive formula for q_n, the number of dots in the nth centered square number.
8. Find an explicit formula for q_n, the number of dots in the nth centered square number.

Catalan Capers

Eugene Charles Catalan (1814–1894), a Belgian mathematician, had a special sequence named for him even though he did not discover the sequence. He did, however, establish results about it. It is Leonhard Euler (1707–1783) who is credited with finding the sequence about 100 years before Catalan's work.

Call the Catalan numbers C_n for $n = 1, 2, 3$, and so on. $C_1 = 1$ and $C_2 = 2$. Name the Catalan numbers C_3 and C_4 by solving this problem.

In how many ways can a convex polygon be dissected into nonoverlapping triangles by drawing diagonals that do not intersect?

Let C_n represent the number of ways a convex polygon with $n + 2$ sides can be dissected into nonoverlapping triangles. Set $C_1 = 1$.

This figure shows how to find $C_2 = 2$.
Quadrilateral

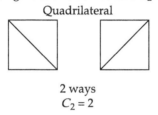

2 ways
$C_2 = 2$

1. Find C_3 by sketching the number of different ways a convex pentagon can be dissected into nonoverlapping triangles. Then, carefully study the examples done for you, and in each case label the diagonals and the bottom side of the pentagon with a matching expression. (Hint: you will not need to use all the pentagons. All expressions should read *a b c d* with parentheses inserted appropriately.)

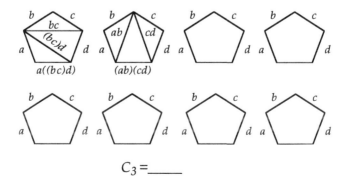

$C_3 =$ _____

2. Find C_4 by sketching the number of different ways a convex hexagon can be dissected into nonoverlapping triangles. Study the example done for you and in each case label the diagonals and bottom side of the hexagon with a matching expression. (Hint: you will not need to use all the hexagons. All expressions should read *a b c d e* with parentheses inserted appropriately.)

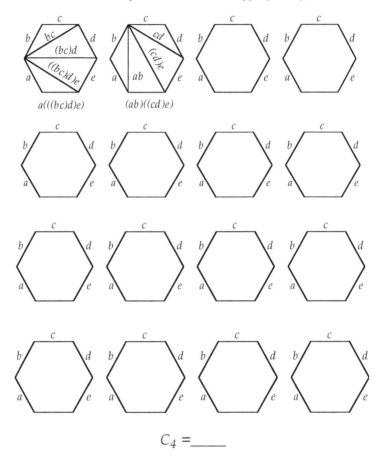

$$C_4 = \underline{\quad\quad}$$

Highly Composite Numbers

A highly composite positive integer $n > 1$ is a positive integer that has more divisors than any positive integer less than n.

Srinivasa Ramanujan (1887–1920), an outstanding number theorist from India, worked with all kinds of number patterns. When the accomplished British mathematician Godfrey Hardy (1877–1947) rated himself and some of his colleagues on a scale of math ability from 0 to 100, he gave himself a score of 25 and Ramanujan a 100. Highly composite numbers were one of Ramanujan's inventions.

The number 2 is highly composite because it is the first number to have *two* divisors, namely 1, 2. Two is the only prime highly composite number.

The number 3 is not highly composite because it does not have more than two divisors.

The number 4 is highly composite because it is the first number to have *three* divisors, namely 1, 2, 4.
1. Name highly composite numbers less than

     a) 25   b) 50

To generate a list of highly composite numbers, it is helpful to use the following fact that counts the number of divisors of a number.

If $N = p_1^{a_1} p_2^{a_2} \times \cdots \times p_k^{a_k}$ then N has $(a_1 + 1)(a_2 + 1) \ldots (a_k + 1)$ divisors. The p_is are primes, and the a_is are the corresponding powers.

For example, consider 30 and 45.
$30 = 2^1 \times 3^1 \times 5^1$, so 30 has $(1 + 1)(1 + 1)(1 + 1) = 8$ divisors.
$45 = 3^2 \times 5$, so 45 has $(2 + 1)(1 + 1) = 6$ divisors.
2. Find the number of divisors of
     a) 60   b) 144   c) 512
3. Determine if any of the following numbers are highly composite:
     a) 100   b) 180   c) 250
4. Use the fact above to find the smallest number with the same number of divisors as 1000.
5. Find out about Ramanujan. What are some of his other discoveries? Write a brief report that describes his life in India, his time in the United Kingdom, and some of the problems he worked on.

The Tower of Hanoi and the Reve's Puzzle

In 1883 Edouard Lucas, a French mathematician who worked with Fibonacci-like sequences and perfect numbers, introduced (under the pseudonym N. Claus) the puzzle he called the *Tower of Hanoi*. Today this puzzle is known under a variety of names and stories to many who are interested in mathematics and puzzles. However, the basic idea of the puzzle is constant. You have a pyramid of n disks as shown below. You must transfer the pyramid from one of three spikes to another using the least number of moves.

The conditions are as follows:
a) you transfer 1 disk per move.
b) you may not place a larger disk on top of a smaller disk.

1. Complete the table:

Number of Disks	1	2	3	4	5	6	7	8	9	10	n
Least No. of Moves	1	3									

2. Explain how

   a) moving 4 disks can be described recursively in terms of the result of moving 3 disks.

   b) moving 10 disks can be described recursively in terms of the result of moving 9 disks.

   c) moving n disks can be described recursively in terms of the result of moving $(n - 1)$ disks.

3. If it takes 1 second to move 1 disk, how long does it take to move a pyramid of

   a) 10 disks? Give your result in minutes.
   b) 15 disks? Give your result in hours.
   c) 20 disks? Give your result in days.
   d) 30 disks? Give your result in years (let 1 year be 365 days).
   e) 64 disks? Give your result in years.

In 1907 the well-known English puzzlist Henry Dudeney published **The Canterbury Puzzles.** This collection of problems based on Geoffrey Chaucer's **Canterbury Tales** contained the Reve's Puzzle.

In Dudeney's narrative the reve asked the traveler: If a stack of eight cheeses graduated in size and arranged top to bottom from smallest to largest were placed on one of four stools, what is the least number of moves needed to transfer the stack of cheeses to another stool?

The rules require that:

a) you transfer 1 cheese per move.
b) you may not place a larger cheese on top of a smaller cheese.

Thus, this puzzle is an extension of the Tower of Hanoi puzzle from three spikes to four spikes.

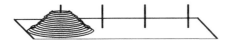

Use the rules described above to determine the following solutions.

4. Complete the table:

Number of Cheeses	1	2	3	4	5	6	7	8
Least No. of Moves								

5. Can you describe your answer for 4 cheeses recursively in terms of your answer for 3 cheeses? How about your answer for 5 cheeses recursively in terms of your answer for 4 cheeses?
6. Dudeney asked for solutions for 8, 10, and 21 cheeses. Using the table can you determine solutions for the cases of 10 and 21 cheeses?

Perfect Number Patterns

> A *perfect number* is a positive integer that equals the sum of its proper divisors. A proper divisor of a number is any divisor of the number excepting itself. The first perfect number is 6. $6 = 1 + 2 + 3$

1. All perfect numbers found to date are even and can be expressed by a formula given by Euclid: $2^{n-1}(2^n - 1)$, where $2^n - 1$ is prime. The formula does not produce a perfect number for all n. Determine which values of $n = 2, 3, 4, 5, 6, 7, 8, 9, 10, 11, 12$ give perfect numbers. List the corresponding perfect numbers.

The number of digits in a number is 1 more than the characteristic in the common log ($\log_{10}$) expression of the number.
Recall, $\log a^b = b \log a$. Use $\log 2 = 0.30103$.

EXAMPLE

Find the number of digits in 2^5.
$\log 2^5 = 5 \log 2 = 5(0.30103) = \underline{1}.50515$

The characteristic is the whole number part of the result, here 1. The number of digits in 2^5 (or 32) is $1 + 1 = 2$.

The number of digits in $2^{n-1}(2^n - 1)$ and in $2^{n-1} \times 2^n$ is the same.

EXAMPLE

Find the number of digits in $2^4(2^5 - 1)$.

$$\log 2^4 \times 2^5 = 9 \log 2 = 9(0.30103) = \underline{2}.70927$$

Since the characteristic is 2, the number of digits is $2 + 1 = 3$.

2. Use a calculator to determine the number of digits in the following perfect numbers:

   a) 10th ($n = 89$)   b) 15th ($n = 1279$)
   c) 20th ($n = 4423$)   d) 25th ($n = 21701$)
   e) 30th ($n = 132049$)

3. The perfect numbers 6 and 28 can be written as sums of powers of 2.
   $6 = 2 + 4$ $28 = 4 + 8 + 16$
   Write 496 and 8128 as sums of powers of 2.

4. Use the fact that the geometric series
   $1 + 2 + 2^2 + \cdots + 2^{n-1} = 2^n - 1$ to show that all perfect numbers of the form $2^{n-1}(2^n - 1)$ are sums of powers of 2.

5. Use a number theory resource or the Internet to learn more about perfect numbers. Write a brief paper that describes and explains other facts about perfect numbers.

Crisscross Cubes

Cube models can be a useful way to physically represent a variety of number sequences. Studying comparisons between successive models leads to the ability to visualize recursive patterns in which new terms can be described using previous term(s). Construct and analyze these cube models.

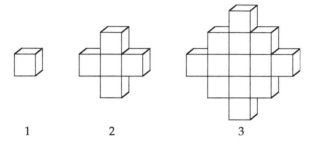

| 1 | 2 | 3 |

1. Build the next three models that follow in the sequence.
2. Use the models to gather data. Complete the table.

Model	Perimeter of Front Face in Units	Volume in Cubic Units	Total Surface Area in Square Units
1	$P_1 = 4$	$V_1 = 1$	$S_1 = 6$
2	$P_2 =$	$V_2 =$	$S_2 =$
3	$P_3 =$	$V_3 =$	$S_3 =$
4	$P_4 =$	$V_4 =$	$S_4 =$
5	$P_5 =$	$V_5 =$	$S_5 =$
6	$P_6 =$	$V_6 =$	$S_6 =$

3. State the generalization for the perimeter of the front face, volume, and total surface area in two ways, namely,

   a) Give a recursive relation for the nth term that uses the preceding term or terms of the sequence.

   b) Give an explicit relation for the nth term in terms of n.

A Medieval Pattern

The number of unit squares in each figure is listed under the figure.

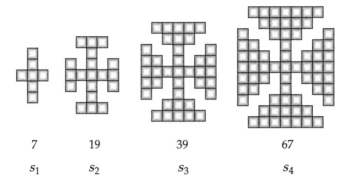

7	19	39	67
s_1	s_2	s_3	s_4

1. Use grid paper and shade in the next figure in the sequence and give the number of unit squares, s_5, in the figure.
2. Complete the table below by listing the next five terms of the sequence s_n.

n	1	2	3	4	5	6	7	8	9	10
Unit Square Count s_n	7	19	39	67						403

3. The terms, s_n, in the table are odd numbers. Explain in a sentence or two why the pattern produces only odd numbers of unit squares.
4. Find a recursive formula for the unit square count, s_n.
5. Find an explicit formula for the unit square count, s_n.
6. Create your own medieval pattern using unit squares. Let s_n be the unit square count.

   a) Draw the first four figures on grid paper.
   b) Prepare a table that displays the first ten terms of your sequence s_n.
   c) Give a recursive or explicit formula for s_n.

Prime Magic

A *magic square* is a square array of numbers in which the sum of the numbers in any row, column, and diagonal is the same number.

The square array

4	14	12
18	10	2
8	6	16

is a magic square with magic constant 30.

The numbers from the array arranged from smallest to largest are

$$2, 4, 6, 8, \underline{10}, 12, 14, 16, 18.$$

Adjacent pairs in the list have a common difference of 2. Also note that 3 times the middle number is the magic constant. This magic square contains exactly one prime number, namely 2. A prime number has exactly two divisors, one and itself.

If a sequence of nine numbers is listed in order of size: $a, b, c, d, e, f, g, h, i$, and the adjacent pairs have the same common difference, then a magic square results when the numbers are arranged as in this 3×3 array.

b	g	f
i	e	a
d	c	h

1. Use this array to find magic squares in which there are exactly 2, 3, 4, 5, 6, 7, and 8 different prime numbers.
2. Find out more facts about magic squares. Write a report to share with the class.

Dealing with Digits (base ten)

> A number is represented by a string of digits. In *base ten*, the possible digits are 0, 1, 2, 3, ... , 9.

1. How many base ten numbers in the interval 1 through 1000 inclusive, have at least one 6?

2. A printer is typesetting a book that has 964 pages. How many of each number occur in the total page numberings? Explain your answers.

   a) 4s   b) 6s   c) 9s

3. A printer analyzes the typesetting for a book and realizes that 1056 digits will be used to number the pages. The book will also have 5 unnumbered pages. If the numbered pages start with 1, how many pages are in the complete book? Describe how you solve the problem.

4. How many three-digit numbers in the range 100 through 999 have one digit that is the average of the other two? For example, in 357 or 573, 5 is the average of 7 and 3. Explain in detail how you get your total.

5. Each arrangement of the digits 1, 2, 3, 4, 5, 6 yields a six-digit number. If the numbers corresponding to all possible arrangements are listed from the smallest (123,456) to the largest (654,321), which number is 413th on the list? Explain in detail how you get your answer.

6. The number names for 10^6, 10^9, and 10^{12} are one million, one billion, and one trillion, respectively. Find number names for higher powers of 10 with exponents that are multiples of 3. Write down the resources you used in finding the names.

7. Consult a history of mathematics book such as G.G. Joseph's **The Crest of the Peacock: Non-European Roots of Mathematics** and write a brief report on the origin of the base ten system of numeration.

Factorial Finishes

> $n!$ is an abbreviation for the product of
>
> $$1 \times 2 \times 3 \times 4 \times \cdots \times n.$$
>
> $n!$ can also be written recursively as
>
> $$n! = n(n - 1)!$$
>
> while $0! = 1$, by definition.

1. $5! = 12\underline{0}$ and 5! ends in one zero. Determine the number of zeros at the end of each of the following. Explain how you arrived at your answer.

   a) 50!   b) 100!   c) 500!

2. The last nonzero digit of $5! = 1\underline{2}0$ is 2. Determine the last nonzero digit of each of the following. Explain how you arrived at your answer.

   a) 20!   b) 25!   c) 30!

3. Produce a list of factorials on a spreadsheet. At what value of n does the answer become approximate? At what value of n does $n!$ become approximate on your calculator?

4. Read about the historical origin of mathematical symbols to find out who was the first to use the factorial symbol. Write a brief report of your findings.

5. Use factorials to explain why there are 15,120 different sequences for the first five moves in the game of tic-tac-toe.

6. 12 girls and 13 boys are planning to line up at the ticket window to purchase football tickets.

   a) How many different lines can be formed?

   b) How many different lines can be formed if all the girls are in front of the boys? Explain your answer.

Designing Designs

Lady Shikibu Murasaki (c. 978–1031) wrote the **Tale of Genji** around 1000 A.D. Because of her literary expertise, she is sometimes called the Shakespeare of Japan. Two chapters in the **Tale of Genji** have *no* design heading, but each of the remaining chapters has a *different* design heading. These designs show the possible arrangements of five separate incense sticks vertically placed side by side that are of the same color or a combination of up to and including five different colors, allowing repetitions. Your challenge, should you decide to accept it, is to determine how many chapters are in the **Tale of Genji**!

Examine a simple case first. You may want to use colored toothpicks to model the arrangements. For example, suppose there are two sticks and up to and including two colors. There are *two* different arrangements of colors:

1 1	1 2	2 1
One 2-stick array in which a *single* color is used. It does not matter which color is chosen.	One 2-stick array in which *two* colors are used.	This 2-stick array is the *same* as the one to the left, since it does not matter which color comes first.

If there are three sticks and up to and including three colors, there are *five* different arrangements of colors.

1 1 1	1 1 2 1 2 1 2 1 1	1 2 3
One 3-stick array in which a *single* color is used. It does not matter which color is chosen.	Three 3-stick arrays in which *two* colors are used. It does not matter which two of the three colors are chosen. It does not matter which color appears twice; it is the position that is important.	One 3-stick array in which *three* colors are used. It does not matter how the three colors are arranged.

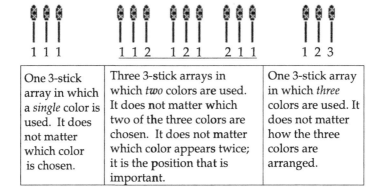

1. Find the 15 different arrangements using four sticks and up to and including four colors. Organize your list according to the number of colors needed.

2. Now you are prepared to try the challenge using five sticks and up to and including five colors.

Fibonacci Fascinations

1. Using a spreadsheet, compile a column of the first 40 Fibonacci numbers
   $F_1, F_2, F_3, \ldots, F_{40}$
   where $F_1 = F_2 = 1$, $F_n = F_{n-1} + F_{n-2}$, and $n > 2$.
2. Develop the following columns based on your column of Fibonacci numbers.
   - CS: The cumulative sum of Fibonacci numbers. The entry in row four of this column should be the sum of

$$1 + 1 + 2 + 3, \quad \text{or} \quad 7.$$

   - RD2: Term remainders after division of Fibonacci numbers by 2. The entry in row four of this column should be 1 since:

$$F_4 \div 2 = 3 \div 2 = 1 \text{ remainder } 1.$$

   - RD3: Term remainders after division of Fibonacci numbers by 3. The entry in row four of this column should be 0 since:

$$F_4 \div 3 = 3 \div 3 = 1 \text{ remainder } 0.$$

   - RD5: Term remainders after division of Fibonacci numbers by 5. The entry in row four of this column should be 3 since:

$$F_4 \div 5 = 3 \div 5 = 0 \text{ remainder } 3.$$

   - Ratio L/S: The ratios of the terms F_n/F_{n-1}
   - Ratio S/L: The ratios of the terms F_{n-1}/F_n
3. Use the data in problem 2. For each column in the spreadsheet, formulate a conjecture about the corresponding Fibonacci number patterns.
4. Find out about Leonardo of Pisa who lived in the 13th century. Provide some evidence to support the claim that he is sometimes called the solitary flame of mathematical genius in the Middle Ages. Write a brief report of your findings.
5. Learn about the Golden Ratio. Investigate some of the many connections between Fibonacci numbers and the Golden Ratio. Write a brief report detailing some of the connections.

Chapter 2

Numbers Based on Divisors and Proper Divisors

Mathematics is the queen of the sciences and number theory is the queen of mathematics.
Carl Friedrich Gauss (1777–1855)

Do you know that the divisors of a number and its primality are related to cryptography? Several secure cipher systems are based on the fact that large numbers are extremely difficult to factor even when using the most powerful computers and efficient algorithms.

Some details about the divisors of a number and their historical connections are provided below to serve as a catalyst for investigation and further exploration.

Positive Integers

The numbers in the set $\{1, 2, 3, 4, 5, \ldots\}$ are the *positive integers*.
Nearly all the integers used in this book are *positive integers*; so to simplify discussions the word *integers,* unless otherwise indicated, will be used to mean *positive integers*.

Divisors, Multiples, and Proper Divisors

> The *divisors* of an integer are those integers that divide the given integer evenly, that is, with zero remainder. The divisors of an integer are also called *integral divisors, exact divisors* or *factors*. The set of divisors of the integer n is denoted D_n.
>
> In the expression,
> $$a = b \times c$$
> where a, b, and c are integers,
> b and c are *divisors* of a
> and a is a *multiple* of b (or c).

EXAMPLE

Find the divisors of 6.

$6 \div 1 = 6$, remainder 0 $6 \div 2 = 3$, remainder 0
$6 \div 3 = 2$, remainder 0 $6 \div 4 = 1$, remainder 2
$6 \div 5 = 1$, remainder 1 $6 \div 6 = 1$, remainder 0

The divisors of 6 are the integers 1, 2, 3, 6.
Write $D_6 = \{1, 2, 3, 6\}$.

> If the integer itself is excluded from the set of its divisors, the new set is called the set of *proper divisors* of the integer, or the set of *aliquot parts*. The set of proper divisors of the integer n is denoted PD_n.

EXAMPLE

The proper divisors of 6 are 1, 2, and 3. Write $PD_6 = \{1, 2, 3\}$.
Note the integer 1 has no proper divisors.

EXERCISE 1

Find the sets of divisors or proper divisors for each integer.

1. D_8 2. D_{10} 3. D_{20}
4. PD_{25} 5. D_{29} 6. PD_{40}
7. D_{41} 8. PD_{41} 9. PD_{73}

Prime and Composite Numbers

Prime and composite numbers are two very important sets of integers.

> A *prime* number is an integer greater than 1 that has only the numbers 1 and itself as divisors.

EXAMPLE

Find the first four prime numbers.
The first four prime numbers are 2, 3, 5, and 7 because $D_2 = \{1, 2\}$, $D_3 = \{1, 3\}$, $D_5 = \{1, 5\}$, and $D_7 = \{1, 7\}$.

> A *composite* number is an integer greater than 1 that has divisors other than 1 and itself. That is, a composite number is an integer greater than 1 that is not a prime.

EXAMPLE

Find the first four composite numbers.

The first four composite numbers are 4, 6, 8, and 9 because $D_4 = \{1, 2, 4\}$, $D_6 = \{1, 2, 3, 6\}$, $D_8 = \{1, 2, 4, 8\}$, and $D_9 = \{1, 3, 9\}$.

Sieve of Eratosthenes

The most famous method for finding primes was developed around 230 B.C. by Eratosthenes, a Greek who was the tutor of the son of Ptolemy III of Egypt and also the chief librarian at the University of Alexandria. The process he used is known today as the Sieve of Eratosthenes.

Algorithm to find prime numbers

1. Write the integers from 1 to as high as you wish. For the purpose of illustration 1 through 100 are used here.

2. Cross out 1 because it is not a prime. Circle ②, which is the first prime number. Every second number after 2 has been crossed out using (\) because each one is divisible by 2 and is therefore composite. Thus 4, 6, 8, . . . , 100 are crossed out.

3. The next number not crossed out is 3. Circle ③; it is a prime. Cross out every third number after 3, with (/), even if it has already been crossed out using (\). Each such number is divisible by 3 or is a multiple of 3 and is composite. Thus cross out 6, 9, 12, . . . , 99.

4. The next number not crossed out is 5. Circle ⑤; it is a prime. Cross out every fifth number after 5 with (|), even if it has already been crossed out using (\) or (/). Thus, cross out 5, 10, 15, 20, . . . , 100.

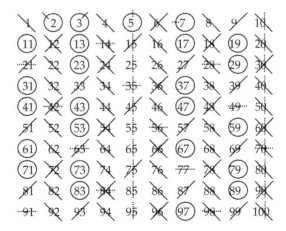

5. Continue in the same way. Circle the next number not crossed out, which is 7. Mark it ⑦ and cross out all the multiples of 7 using (—).

6. Now circle the numbers that remain. They are the primes less than 100.

If you are seeking primes greater than 100, it is necessary to continue the process by crossing out multiples of 11, 13, and so on.

EXERCISE 2

1. For each of the numbers below, give its divisors, its proper divisors, and indicate whether the number is prime or composite.

   a) 18   b) 23   c) 24   d) 28
   e) 31   f) 37   g) 39   h) 59

2. Consider the set of integers greater than 10.
   a) In which digits do the prime numbers appear to end?
   b) In which digits do the composite numbers appear to end?

3. When using the Sieve of Eratosthenes to find prime numbers from 1 to 100, explain why all primes were determined once the multiples of 7 were crossed out.

4. Just as 7 was the last number whose multiples were crossed out to find all primes less than 100, which number is the last one whose multiples are crossed out to find all primes less than

   a) 200?   b) 300?

   Give a reason for your choice.

Prime Factorization Property (Fundamental Theorem of Arithmetic)

The term *factor* connects prime and composite numbers. Recall that a factor of a number is an exact divisor of the number. A factor that is a prime number is called a *prime factor* or a *prime divisor.*

Study this array.

Composite number	Prime factors	Composite number as a product of prime factors
6	2, 3	$6 = 2 \times 3$
8	2	$8 = 2 \times 2 \times 2 = 2^3$
10	2, 5	$10 = 2 \times 5$
12	2, 3	$12 = 2 \times 2 \times 3 = 2^2 \times 3$
100	2, 5	$100 = 2 \times 2 \times 5 \times 5 = 2^2 \times 5^2$

Notice that

a) each composite number has been written as a product of its prime factors, raised to the first or higher powers, and

b) if the order of the prime factors in the product is not considered, then there is exactly *one* way to write the product.

Statements (a) and (b) comprise the *Prime Factorization Property* for composite numbers.

> **Prime Factorization Property:** Every composite number can be factored into a product of primes in exactly one way. This statement is also called the *Fundamental Theorem of Arithmetic.*

How can you find out whether a given number is a prime or a composite? For extremely large numbers, there is no simple procedure. For smaller numbers, the prime factorization of composite numbers provides a way.

EXAMPLE

Is 147 prime? Composite? If 147 is composite, write it as a product of its prime factors.

The procedure is to check to see if any of the primes $2, 3, 5, 7, 11, 13, \ldots$ are factors of 147.

a) Begin with the smallest prime. 2 is not a factor of 147, so try 3.

b) You can divide to determine that 3 is a factor of 147.

$$3 \underline{| \ 147}$$
$$49 \qquad \text{and } 147 = 3 \times 49$$

c) Continue the procedure with 49. It has been determined that 2 is not a factor, so try 3, then 5. But neither one is a factor. The next prime, 7, is a factor of 49.

$$3 \underline{| \ 147}$$
$$7 \underline{| \ 49}$$
$$7$$

d) The last quotient, 7, is a prime, and the search is completed.

e) The product of all the prime factors found by this procedure should equal the given number. The composite number

$$147 = 3 \times 7 \times 7 = 3 \times 7^2$$

EXAMPLE

Is 225 prime? Composite? If 225 is composite, write it as a product of its prime factors.

$$3 \underline{| \ 225}$$
$$3 \underline{| \ 75}$$
$$5 \underline{| \ 25}$$
$$5$$

The composite number $225 = 3 \times 3 \times 5 \times 5 = 3^2 \times 5^2$

EXAMPLE

Is 79 prime? Composite? If 79 is composite, write it as a product of its prime factors.

None of the primes $2, 3, 5, 7, 11, 13, \ldots$ is a factor of 79. Thus 79 is a prime number.

EXERCISE 3

For each of the numbers below, determine whether the number is prime or composite. If the number is composite, write it as a product of its prime factors.

1. 79	2. 84	3. 93	4. 101	5. 114	6. 127
7. 133	8. 144	9. 152	10. 200	11. 263	12. 495

Testing for Primes

How many primes must be tested as possible prime factors in order to find out if a given number is prime, or to get its prime factorization? You may use the following rule as a general guide.

Test as possible prime factors of a given number only those primes that are less than or equal to the square root of the number.

This rule is a consequence of observing:

1) if the distinct factor pairs a, b of $N = ab$ are listed in order of size starting with the smallest a, then there is a point at which repetition of factor pairs a, b occurs in the reverse order b, a. This turning point is $\sqrt{N}$.
2) that from the Sieve of Eratosthenes you need check only for *prime* factors a of N.

You will find, in practice, that it may not be necessary to test all primes up through the square root of the number. Try to find short cuts wherever possible.

EXAMPLE

Find the prime factorization of 100.

$$100 = 10 \times 10 \qquad 10 = 2 \times 5, \quad \text{so } 100 = 2^2 \times 5^2$$

EXAMPLE

Determine whether 73 is a prime number.

Test all primes less than $\sqrt{73}$. Since $\sqrt{73}$ is approximately 8.54, test the primes $2, 3, 5,$ and 7. None of these divides 73. Hence, 73 is a prime number.

EXAMPLE

Determine if 143 is a prime number.

Test all primes less than $\sqrt{143}$. Since $\sqrt{143}$ is approximately 11.96, test the primes 2, 3, 5, 7, and 11. Eleven divides 143 because $143 = 11 \times 13$; so 143 is a composite number.

EXERCISE 4

For each of the numbers below, determine whether the number is prime or composite. If the number is composite, write it as a product of its prime factors.

1. 30	2. 73	3. 96	4. 111	5. 124	6. 196
7. 233	8. 289	9. 323	10. 343	11. 497	12. 551

Divisors of an Integer, GCD, and LCM

Exponential notation is used in order to keep track of all the exact divisors of a given integer.

$$a^n = \underbrace{a \times a \times a \times a \times \cdots \times a}_{n \text{ factors}}$$

where n is an integer, and $a \neq 0$, and $a^0 = 1$.

All the exact divisors of an integer can be found easily from the prime factorization of the integer. The divisors can be listed quickly using power notation. Examine the table below very carefully. Find patterns for the prime factors and for the exponents.

Number	512	6	12	36
Prime Factorization	2^9	$2^1 \times 3^1$	$2^2 \times 3^1$	$2^2 \times 3^2$
Divisors of Number	$2^0 = 1$	$2^0 = 1$	$2^0 = 1$	$2^0 = 1$
	$2^1 = 2$	$2^1 = 2$	$2^1 = 2$	$2^1 = 2$
	$2^2 = 4$	$3^1 = 3$	$3^1 = 3$	$3^1 = 3$
	$2^3 = 8$	$2^1 \times 3^1 = 6$	$2^2 = 4$	$2^2 = 4$
	$2^4 = 16$		$2^1 \times 3^1 = 6$	$3^2 = 9$
	$2^5 = 32$		$2^2 \times 3^1 = 12$	$2^1 \times 3^1 = 6$
	$2^6 = 64$			$2^2 \times 3^1 = 12$
	$2^7 = 128$			$2^1 \times 3^2 = 18$
	$2^8 = 256$			$2^2 \times 3^2 = 36$
	$2^9 = 512$			

EXERCISE 5

List all of the divisors of each number show below. You may wish to make a table similar to the one above to organize your information.

1. $24 = 2^3 \times 3^1$
2. $72 = 2^3 \times 3^2$
3. $216 = 2^3 \times 3^3$
4. $30 = 2^1 \times 3^1 \times 5^1$
5. $60 = 2^2 \times 3^1 \times 5^1$
6. $180 = 2^2 \times 3^2 \times 5^1$

In the table on p. 45, $512 = 2^9$ and 512 has 10 divisors. Study all the data in the table carefully. Use your findings to calculate the number of divisors of each given number.

7. 15
8. 18
9. 100
10. 72
11. 5^p
12. 2×3^n
13. $a^p \times b^q$, where a and b are prime.

The *greatest common divisor (gcd)* of a pair of integers is the largest integer that divides each of the given numbers. The *least common multiple (lcm)* of a pair of integers is the smallest integer that contains each of the given numbers as a divisor.

EXAMPLE

Use prime factorization to find the gcd and lcm of 84 and 441.

$$84 = 2^2 \times 3 \times 7 \qquad 441 = 3^2 \times 7^2$$

The gcd of 84 and 441 is $3 \times 7 = 21$.
The lcm of 84 and 441 is $2^2 \times 3^2 \times 7^2$.

EXERCISE 6

There is an interesting connection between the gcd and lcm of a pair of numbers.

$$\text{lcm}(a, b) = \frac{a \times b}{\gcd(a, b)}$$

1. Verify the above relation for $a = 54$ and $b = 60$.
2. Use the above relationship to find lcm (40, 28).
3. Use the above relationship to find gcd (36, 123).
4. The gcd of a pair of numbers (a, b) is 6 and their lcm is 90. List the possible replacements for a and b that fit the description.
5. Explain why the above relation is valid.

Try the **Investigations: Factor Lattices, Fermat Factorings, A Juggling Act, and Highly Composite Numbers.**

Relatively Prime and Euler ϕ Numbers

Leonhard Euler (1707–1783) (read: *Oiler*), of Basel, Switzerland, is one of the world's greatest mathematicians. He spent considerable time doing mathematics in St. Petersburg, Russia and Berlin, Germany. Euler was the father of 13 children. He became blind in his later years, yet his loss of sight did not diminish his mathematical productivity. In over 750 books and articles, he made significant contributions to every branch of mathematics.

Euler's ϕ (Greek letter, phi, read: *fi)* numbers use the concept, *relatively prime.*

> Two numbers are *relatively prime* if they have 1 as their greatest common divisor. Relatively prime numbers are also called *co-prime numbers.*

EXAMPLE

The pairs 1 and 3, 3 and 7, 4 and 9, and 8 and 15 are relatively prime numbers.

EXAMPLE

The numbers 2 and 6 are not relatively prime since their greatest common divisor is 2.

An Euler ϕ number, symbolized by $\phi(n)$, is associated with each integer n.

> $\phi(n)$ is the number of numbers greater than 0 and less than or equal to n that are relatively prime to n.
> $\phi(1)$ is defined to be 1.

EXAMPLE

Find $\phi(5)$.

1, 2, 3, and 4 are all relatively prime to 5. So $\phi(5) = 4$. Therefore, 4 is the *Euler ϕ* number for 5.

EXAMPLE

Find $\phi(6)$.

The numbers less than or equal to 6 that are relatively prime to 6 are 1 and 5. So $\phi(6) = 2$, or the Euler ϕ number for 6 is 2.

EXERCISE 7

1. Determine if each of the following pairs of numbers are relatively prime.

a) 1, 5	b) 2, 9	c) 10, 13
d) 12, 16	e) 14, 22	f) 16, 17
g) 21, 41	h) 43, 86	i) 49, 77

j) 52, 100   k) 63, 81   l) 70, 71
m) 75, 145   n) 86, 110   o) 91, 112

2. Are any two different prime numbers relatively prime? Explain.
3. Are any two consecutive integers relatively prime? Explain.
4. For each of the integers from 1 through 14, find $\phi(n)$. List the numbers that are less than or equal to n and relatively prime to n.
5. Find the Euler ϕ number for several prime numbers. Can you predict what the Euler ϕ number will be for a prime number? Explain your answer.
6. Find the Euler ϕ number for several powers of 2. Can you predict what the Euler ϕ number will be for a particular power of 2? Explain your answer.
7. For each pair of integers (a, b) below, compute $\phi(a \times b)$ and $\phi(a) \times \phi(b)$. When and why are the two results equal?

a) (3, 5)   b) (3, 6)   c) (4, 5)   d) (4, 6)   e) (2, 2)
f) (2, 5)   g) (2, 6)   h) (2, 7)   i) (2, 9)   j) (5, 5)

Abundant, Deficient, and Perfect Numbers

Here are some integers whose properties are related to their divisors and proper divisors.

An integer is *abundant* if the sum of its proper divisors is greater than the integer. Abundant numbers are also called *redundant* or *excessive* numbers.

EXAMPLE

Show that 12 is an abundant number.

$$PD_{12} = \{1, 2, 3, 4, 6\} \quad \text{and} \quad 1 + 2 + 3 + 4 + 6 = 16$$

Since 16 is greater than 12, 12 is an abundant number.

An integer is *deficient* if the sum of its proper divisors is less than the integer. Deficient numbers are also called *defective* numbers.

EXAMPLE

Show that 9 is a deficient number.

$$PD_9 = \{1, 3\} \quad \text{and} \quad 1 + 3 = 4$$

Since 4 is less than 9, 9 is a deficient number.

An integer is *perfect* if the sum of its proper divisors is equal to the integer.

EXAMPLE

Show that 6 is a perfect number.

$$PD_6 = \{1, 2, 3\} \quad \text{and} \quad 1 + 2 + 3 = 6$$

The sum 6 equals the integer 6, so 6 is a perfect number.

EXAMPLE

Show that 28 is a perfect number.

$$PD_{28} = \{1, 2, 4, 7, 14\} \quad \text{and} \quad 1 + 2 + 4 + 7 + 14 = 28$$

The sum 28 equals the integer 28, so 28 is a perfect number.

EXERCISE 8

Determine whether each number is abundant or deficient. First list the proper divisors of the number and then find their sum.

1. 8	2. 20	3. 23	4. 24	5. 30	6. 56
7. 60	8. 79	9. 100	10. 148	11. 155	12. 186

EXERCISE 9

1. Compile a list of the first 50 abundant numbers. Are there any odd abundant numbers in your list? If so, name them.

2. a) Is 945 an abundant number? Explain your choice.
   b) Is 1575 an abundant number? Explain your choice.

Sums and Differences of Abundant and Deficient Numbers

What kinds of numbers result from performing the operations of addition and subtraction on abundant and deficient numbers?

EXAMPLE

Show that the sum of abundant numbers may be abundant.
Consider the sum of the two abundant numbers 12 and 18.

$$12 + 18 = 30$$
$$PD_{30} = \{1, 2, 3, 5, 6, 10, 15\}$$
$$1 + 2 + 3 + 5 + 6 + 10 + 15 = 42$$

The sum 42 is greater than the integer 30, so in this case the sum of two abundant numbers is also an abundant number.

EXERCISE 10

1. Each of the indicated sums below is the sum of two abundant numbers. First compute the sum, then test to see if the sum is an abundant number following the procedure in the preceding example.

   a) $12 + 24$ b) $18 + 20$ c) $18 + 30$
   d) $20 + 24$ e) $20 + 30$ f) $24 + 30$

2. Determine if the sum of two abundant numbers is always an abundant number or just sometimes an abundant number. Explain your answer.

EXAMPLE

Create chains of abundant numbers.

Start with the abundant number 24 and add the abundant number 12. Then continue adding 12 to each sum.

24	36	48	60	72	84	96	108
+ 12	+ 12	+ 12	+ 12	+ 12	+ 12	+ 12	+ 12
36	48	60	72	84	96	108	120

Verify that each sum is abundant. Note that each sum is a multiple of 12.

EXERCISE 11

Follow the procedure used in the preceding example to make more chains of numbers. Start with the first number given and add the second number until you have listed 8 sums. Determine whether or not each of the sums listed is an abundant number.

1. Start with 36 and add 20.   2. Start with 48 and add 36.
3. Start with 60 and add 18.   4. Start with 72 and add 60.
5. Start with 84 and add 72.

EXAMPLE

Investigate sums of deficient numbers.
Compute the sum of the deficient numbers 17 and 19.

$$17 + 19 = 36$$
$$PD_{36} = \{1, 2, 3, 4, 6, 9, 12, 18\}$$
$$1 + 2 + 3 + 4 + 6 + 9 + 12 + 18 = 55$$

The sum 55 is greater than the integer 36, so in this case the sum of two deficient numbers is an abundant number.

EXERCISE 12

1. Each of the indicated sums below is the sum of two deficient numbers. First compute the sum, then test to see if the sum is a deficient number, following the

procedure in the preceding example.

a) $15 + 17$ b) $16 + 22$ c) $21 + 23$

d) $27 + 32$ e) $26 + 38$ f) $35 + 37$

2. Is the sum of two deficient numbers always an abundant number or just sometimes an abundant number?
   Explain your reasoning.

EXAMPLE

Try sums of an abundant number and a deficient number.

Consider the sum of the abundant number 18 and the deficient number 22.

$$18 + 22 = 40$$

$$PD_{40} = \{1, 2, 4, 5, 8, 10, 20\}$$

$$1 + 2 + 4 + 5 + 8 + 10 + 20 = 50$$

The sum 50 is greater than the integer 40, so in this case the sum of an abundant
number and a deficient number is abundant.

EXERCISE 13

1. Each of the indicated sums below is the sum of an abundant number and a
   deficient number. First compute the sum; then test to see if the sum is abundant,
   following the procedure in the preceding example.

a) $12 + 14$ b) $18 + 16$ c) $20 + 16$

d) $24 + 32$ e) $30 + 14$ f) $36 + 22$

2. Is the sum of an abundant number and a deficient number always abundant or
   just sometimes abundant?
   Explain your reasoning.

EXAMPLE

Explore differences of two abundant numbers.

Consider the difference of the abundant numbers 30 and 18.

$$30 - 18 = 12$$

$$PD_{12} = \{1, 2, 3, 4, 6\}$$

$$1 + 2 + 3 + 4 + 6 = 16$$

The sum 16 is greater than the integer 12, so in this case the difference of two
abundant numbers is abundant.

EXERCISE 14

1. Each indicated difference on p. 52 is the difference of two abundant numbers.
   First compute the difference. Then, test to see if the difference is abundant,

following the procedure in the preceding example.

a) $48 - 30$ b) $40 - 30$ c) $70 - 40$

d) $78 - 70$ e) $56 - 40$ f) $100 - 56$

2. Is the difference of two abundant numbers always abundant or just sometimes abundant? Explain your reasoning.

EXAMPLE

Examine differences of two deficient numbers.

Consider the difference of the deficient numbers 37 and 23.

$$37 - 23 = 14$$
$$PD_{14} = \{1, 2, 7\}$$
$$1 + 2 + 7 = 10$$

The sum 10 is less than the integer 14, so in this case the difference of two deficient numbers is deficient.

EXERCISE 15

1. Each indicated difference on p. 53 is the difference of two deficient numbers. First compute the difference. Then test it to see if the difference is deficient.

a) $38 - 11$ b) $22 - 14$ c) $33 - 15$

d) $44 - 11$ e) $64 - 16$

2. Is the difference of two deficient numbers always deficient or just sometimes deficient? Explain your reasoning.

EXAMPLE

Try differences of an abundant number and a deficient number. Consider the difference of the abundant number 60 and the deficient number 22.

$$60 - 22 = 38$$
$$PD_{38} = \{1, 2, 19\}$$
$$1 + 2 + 19 = 22$$

The sum 22 is less than the integer 38, so in this case, the difference of an abundant number and a deficient number is deficient.

EXERCISE 16

1. Each indicated difference on p. 53 is the difference of an abundant number and a deficient number. First compute the difference. Then test it to see if the

difference is abundant or deficient.

a) $48 - 15$	b) $40 - 10$	c) $96 - 47$
d) $138 - 31$	e) $186 - 46$	f) $208 - 96$
g) $33 - 12$	h) $52 - 12$	i) $117 - 78$
j) $158 - 104$	k) $188 - 84$	

2. Is the difference of an abundant number and a deficient number always deficient or just sometimes deficient? Explain your answer.

Products of Abundant and Deficient Numbers

There are three types of products to consider:

$$abundant \times abundant$$
$$abundant \times deficient$$
$$deficient \times deficient$$

EXAMPLE

Is the product of the abundant number 12 and the deficient number 4 abundant?

$$12 \times 4 = 48$$
$$PD_{48} = \{1, 2, 3, 4, 6, 8, 12, 16, 24\}$$
$$1 + 2 + 3 + 4 + 6 + 8 + 12 + 16 + 24 = 76$$

The sum 76 is greater than the integer 48, so here the product of an abundant and a deficient number is abundant.

EXERCISE 17

1. Each indicated product below is the product of two abundant numbers. First compute the product. Then, test to see whether or not the product is abundant.

a) 12×18	b) 12×24	c) 12×30	d) 12×42
e) 18×18	f) 18×24	g) 18×30	h) 18×36
i) 24×36	j) 24×42		

2. In problem 1, does the product of two abundant numbers appear to be always abundant or just sometimes abundant? Explain your answer.

3. Each indicated product below is the product of an abundant number and a deficient number. First compute the product. Then, test to see whether or not the product is abundant.

a) 3×48	b) 4×20	c) 5×36	d) 7×12
e) 7×42	f) 9×42	g) 11×18	h) 11×36
i) 13×18	j) 13×24		

4. In problem 3, is the product of an abundant number and a deficient number always abundant or just sometimes abundant? Explain your answer.

5. Which of the labels, finite or infinite, better describes the set of abundant numbers? Explain your answer.

6. Each indicated product below is the product of two deficient numbers. First compute the product. Then, test to see whether or not the product is deficient.

   a) 11×17 b) 11×31 c) 14×15 d) 14×17
   e) 15×22 f) 16×21 g) 17×31 h) 19×19

7. Is the product of two deficient numbers always deficient or just sometimes deficient? Explain your answer.

Multiples of Perfect Numbers

Recall that an integer is perfect if the sum of its proper divisors is equal to the integer itself.

EXAMPLE

Consider the perfect number 28. Decide if $5 \times 28 = 140$ is abundant. The proper divisors of 140 include all of the divisors of 28 as well as several new numbers that contain 5 as a factor.

$$PD_{28} = \{1, 2, 4, 7, 14\}$$

$$1 + 2 + 4 + 7 + 14 = 28$$

$$PD_{140} = \{1, 2, 4, 7, 14, 28, 5 \times 1, 5 \times 2, 5 \times 4, 5 \times 7, 5 \times 14\}$$

$$\underbrace{1 + 2 + 4 + 7 + 14 + 28} \quad + 5 + 10 + 20 + 35 + 70$$

$$= 2 \times 28 \qquad\qquad\qquad + 5 + 10 + 20 + 35 + 70$$

$$= 56 + 140 = 196$$

The sum of the proper divisors of 140, 196 is greater than the integer 140, so this multiple of a perfect number is abundant.

EXERCISE 18

1. Compute the following multiples of the perfect number 6.

   a) 2×6 b) 3×6 c) 7×6 d) 8×6 e) 10×6

2. Are multiples of 6 always abundant or just sometimes abundant? Explain.

3. Compute the following multiples of the perfect number 28. Determine whether or not each multiple is abundant.

   a) 2×28 b) 3×28 c) 7×28 d) 8×28 e) 10×28

4. Are multiples of 28 always abundant or just sometimes abundant? Explain.

Consecutive Integers and Abundant Numbers

The number pairs (1, 2), (3, 4), and (12, 13) are examples of pairs of consecutive integers. Note in each pair that the second integer is one more than the first integer. The number triplets (1, 2, 3), (10, 11, 12), and (22, 23, 24) are examples of three consecutive integers. The third integer is one more than the second and the second integer is one more than the first in each triplet. Sets of four or more consecutive integers can be described in a similar way.

Are products of consecutive integers abundant numbers?

EXAMPLE

Consider the products of the consecutive pairs (3, 4), (4, 5), and (5, 6).

$$3 \times 4 = 12 \qquad 4 \times 5 = 20 \qquad 5 \times 6 = 30$$

The products 12, 20, and 30 are all abundant numbers.

EXERCISE 19

1. Compute the product of each of the following pairs of consecutive integers. Determine whether or not each product is abundant.

a) 8×9	b) 9×10	c) 10×11	d) 11×12
e) 12×13	f) 13×14	g) 14×15	h) 15×16
i) 16×17	j) 17×18	k) 19×20	l) 22×23

2. Is the product of two consecutive integers always abundant or just sometimes abundant? Explain your answer.

3. Compute the product of each of the following triplets of consecutive integers. Determine whether or not each product is abundant. If possible, write each product as a multiple of 6.

a) $2 \times 3 \times 4$	b) $3 \times 4 \times 5$	c) $4 \times 5 \times 6$
d) $5 \times 6 \times 7$	e) $6 \times 7 \times 8$	f) $8 \times 9 \times 10$

4. Is the product of three consecutive integers greater than 1 always abundant or just sometimes abundant? Explain your answer.

5. Write each product below as a multiple of 6.

a) $2 \times 3 \times 4 \times 5$	b) $7 \times 8 \times 9 \times 10$
c) $8 \times 9 \times 10 \times 11 \times 12$	d) $13 \times 14 \times 15 \times 16 \times 17$

6. a) Is it possible to write the product of any four or more consecutive integers as a multiple of 6? Explain why or why not.

   b) Are all products of four or more consecutive integers abundant? Explain why or why not.

Abundant Numbers as Sums of Abundant Numbers

Some abundant numbers can be written as sums of other abundant numbers. Some can be written in just one way; others can be written in several different ways.

EXAMPLE

Write 24 as the sum of two abundant numbers.

$$24 = 12 + 12$$

There is exactly one way to write 24 as the sum of two abundant numbers. The smallest abundant number that can be written as the sum of two abundant numbers in exactly one way is 24.

EXAMPLE

Write 36 as the sum of two abundant numbers.

$$36 = 12 + 24 = 18 + 18$$

Thirty-six is the smallest abundant number that can be written as the sum of two abundant numbers in exactly two ways.

EXERCISE 20

1. Find the smallest abundant number that can be written as the sum of *two* abundant numbers in exactly the number of ways requested.

   a) three ways   b) four ways   c) five ways   d) six ways
   e) seven ways   f) eight ways   g) nine ways

2. Find the smallest abundant number that can be written as the sum of *three* abundant numbers in exactly the number of ways requested.

   a) one way   b) two ways   c) three ways   d) five ways
   e) six ways   f) eight ways   g) nine ways

Powers of Primes and Deficient Numbers

Study the following table. It appears that the sum of the proper divisors of an integral power of 2 is 1 less than the number itself.

Number: Power of 2	Proper divisors	Sum of proper divisors	Sum = number − 1
$2^1 = 2$	1	1	$1 = 2^1 - 1$
$2^2 = 4$	1, 2	3	$3 = 2^2 - 1$
$2^3 = 8$	1, 2, 4	7	$7 = 2^3 - 1$

EXERCISE 21

1. Extend the pattern in the table above for 2^4 through 2^{10}.
2. If the sum of the proper divisors of each power of 2 can be shown to be 1 less than the number itself, then will each power of 2 be deficient? Explain your answer.
3. a) What number is the only proper divisor of a prime number?
   b) Is the sum of the proper divisors of any prime number greater or less than the number?
   c) Do all prime numbers belong in the set of abundant numbers? Deficient numbers? Perfect numbers?
   d) Since it can be shown that the number of primes is infinite, what other set of numbers is also infinite?

Since the powers of the prime 2 are deficient, do you think the powers of other primes are deficient?

Power of 3	Proper divisors	Sum of proper divisors	Deficient?
$3^1 = 3$	1	1	yes
$3^2 = 9$	1, 3	4	yes
$3^3 = 27$	1, 3, 9	13	yes

The next example illustrates a shortcut method for finding the sum of the proper divisors of a power of any prime number. The sum for any prime p raised to the power n can be shown to be

$$\frac{p^n - 1}{p - 1}$$

EXAMPLE

Verify that the sum of the proper divisors of the powers of 3, 5, and 7 are the numbers given by the formula.

sum of $PD_{3^1} = 1$ $\qquad \dfrac{3^1 - 1}{3 - 1} = \dfrac{2}{2} = 1$

sum of $PD_{5^3} = 1 + 5 + 25 = 31$ $\qquad \dfrac{5^3 - 1}{5 - 1} = \dfrac{124}{4} = 31$

sum of $PD_{7^2} = 1 + 7 = 8$ $\qquad \dfrac{7^2 - 1}{7 - 1} = \dfrac{48}{6} = 8$

Notice that 3^1, 5^3, and 7^2 are all deficient.

EXERCISE 22

1. Determine whether or not each of the powers of primes below are deficient. Use the shortcut method illustrated in the example on p. 57 to compute the sum of the proper divisors.

   a) 3^4 b) 7^4 c) 11^2 d) 11^3 e) 13^2

   f) 17^2 g) 19^3 h) 23^2 i) 29^2

2. Which powers of primes in problem 1 are deficient? Formulate a conjecture, a statement that you expect might be true.

EXAMPLE

Consider the following products of primes.

$2 \times 3 = 6$ $2 \times 13 = 26$ $11 \times 17 = 187$ $2 \times 41 = 82$

The first product, 6, is perfect. The others are deficient numbers.

EXERCISE 23

1. Determine whether or not each product is deficient.

   a) 3×5 b) 3×7 c) 5×11 d) 7×13 e) 7×17

   f) 11×17 g) 17×19 h) 23×29 i) 11×23 j) 17×23

2. Which products in problem 1 are not deficient? Formulate a conjecture.

Even and Odd Integers, Even Perfect Numbers, Mersenne Primes

An *even* integer is an integer that has a remainder 0 when divided by 2. The nonnegative integers 0, 2, 4, 6, 8, ... are even.

An *odd* integer is an integer that has a remainder 1 when divided by 2. The integers 1, 3, 5, 7, 9, ... are odd.

EXERCISE 24

1. a) In which digits must all even integers end?
   b) In which digits must all odd integers end?

2. Construct several examples of sums and products of even and odd numbers. Look for generalizations you can make about the sums and products. Summarize your results in the tables below.

Addition		
+	even	odd
even	even	
odd		

Multiplication		
×	even	odd
even	even	
odd		

3. a) $PD_{496} = \{1, 2, 4, 8, 16, 31, 62, 124, 248\}$. Give the sum of the proper divisors of 496.

   b) $PD_{8128} = \{1, 2, 4, 8, 16, 32, 64, 127, 254, 508, 1016, 2032, 4064\}$. Give the sum of the proper divisors of 8128.

4. Each of the products below can be written in the form $2^{n-1}(2^n - 1)$, where n is a positive integer. First write each product in that form. Then determine whether or not the product is both even and perfect. Problems (a) and (b) have been started for you.

   a) $1 \times 1 = 2^0(2^1 - 1)$. Product neither even nor perfect.

   b) $2 \times 3 = 2^1(2^2 - 1)$ c) 4×7 d) 8×15

   e) 6×31 f) 32×63 g) 64×127

Try the **Investigation: The Tower of Hanoi and The Reve's Puzzle.**

The mathematician René Descartes in 1637 said "Perfect numbers like perfect persons are very rare." Perfect numbers have a rich history that can be traced back at least to Euclid.

Euclid (c. 300 B.C.) was the author of the well-known and highly respected collection of 13 books, entitled the **Elements** *(Stoicheia)*, which included topics in geometry and number theory. Euclid knew about even perfect numbers.

Even perfect numbers are the product of a power of 2 and a special prime number called a *Mersenne prime*. That is, even perfect numbers can be written in the form

$$2^{n-1} (2^n - 1), \text{ with } n \text{ a prime number.}$$

$$\uparrow \qquad \uparrow$$

power of 2   *Mersenne prime*

Marin Mersenne (1588–1648) was a French Franciscan monk who regularly corresponded with several well-known mathematicians and intellectuals of his time. Among his interests were numbers of the form $2^n - 1$ that were prime. The first four Mersenne primes are 3, 7, 31, and 127.

As of the present writing, there are only 48 known even perfect numbers and 48 Mersenne primes. No odd perfect numbers have been found, but no one has been able to prove that they do not exist.

The Great Internet Mersenne Prime Search (GIMPS) is a global organization, founded in 1996 by George Woltman, of thousands of volunteers, experts and amateurs alike, who use their computers and specially designed software, Prime95, to search for Mersenne primes. As of February 2013, the GIMPS project had found 14 Mersenne primes. In addition to seeking perfect numbers larger than the 48th, the search is ongoing to ensure that no additional perfect numbers have been overlooked in the gaps between perfect numbers listed from the 43rd through the 48th perfect numbers. Consult the Internet for additional details of this supercomputing endeavor.

The following is a list of the 48 known even perfect numbers with their digit count, date of discovery and discoverer(s).

Try the **Investigation: Perfect Number Patterns**.

Perfect Numbers of the Form $2^{n-1}(2^n - 1)$

	Prime n	Number of Digits	Date	Discoverer(s)
1.	2	1	Unknown	Known to Greeks
2.	3	2	Unknown	Known to Greeks
3.	5	3	Unknown	Known to Greeks
4.	7	4	Unknown	Known to Greeks
5.	13	8	1456	Recorded in Codex 14908*
6.	17	10	1588	Cataldi
7.	19	12	1588	Cataldi
8.	31	19	1772	Euler
9.	61	37	1883	Pervushin
10.	89	54	1911	Powers
11.	107	65	1914	Powers
12.	127	77	1876	Lucas
13.	521	314	1952	Robinson
14.	607	366	1952	Robinson
15.	1279	770	1952	Robinson
16.	2203	1327	1952	Robinson
17.	2281	1373	1952	Robinson
18.	3217	1937	1957	Riesel
19.	4253	2561	1961	Hurwitz
20.	4423	2663	1961	Hurwitz
21.	9689	5834	1963	Gillies
22.	9941	5985	1963	Gillies
23.	11213	6751	1963	Gillies
24.	19937	12,003	1971	Tuckerman
25.	21701	13,066	1978	Noll & Nickel
26.	23209	13,973	1979	Noll
27.	44497	26,790	1979	Nelson & Slowinski
28.	86243	51,924	1983	Slowinski
29.	110503	66,530	1988	Colquitt & Welsh
30.	132049	79,502	1983	Slowinski
31.	216091	130,100	1985	Slowinski
32.	756839	455,663	1992	Slowinski & Gage
33.	859433	517,430	1994	Slowinski & Gage
34.	1257787	757,263	1996	Slowinski & Gage
35.	1398269	841,842	1996	Armengaud**
36.	2976221	1,791,864	1997	Spence
37.	3021377	1,819,050	1998	Clarkson
38.	6972593	4,197,919	1999	Hajratwala
39.	13466917	8,107,892	2001	Cameron
40.	20996011	12,640,858	2003	Shafer

(Continued)

(*Continued*)

	Prime n	Number of Digits	Date	Discoverer(s)
41.	24036583	14,471,465	2004	Findley
42.	25964951	15,632,458	2005	Nowak
43.	30402457	18,304,103	2005	Cooper, Boone
44.	32582657	19,616,714	2006	Cooper, Boone
45.	37156667	22,370,543	2008	Elvenich
46.	42643801	25,674,127	2009	Strindmo
47.	43112609	25,956,377	2008	Smith
48.	57885161	34,850,340	2013	Cooper

*The oldest, undisputed record appears in this codex from a Benedictine monastery (Part 1-1456, Part 2-1461).
**GIMPS began in 1996 with subsequent perfect numbers discovered by individuals through GIMPS using Prime95.

The table above is testimony to the fact that perfect numbers have fascinated research and amateur mathematicians and others for centuries. In his book, **Mathematical Magic Show**, Martin Gardner (1914–2010), well-known writer on recreational mathematics applauded for his popularization of mathematics through his column in the *Scientific American*, made the following observation: "One would be hard put to find a set of whole numbers with a more fascinating history and more elegant properties surrounded by greater depths of mystery — and more totally useless — than the perfect numbers."

Multiply Perfect Numbers

A *multiply perfect* number is an integer, the sum of whose divisors is a multiple of the integer. Multiply perfect numbers are also called *pluperfect* or *multiperfect* numbers. When the sum of all the divisors is twice the number, the number is *doubly perfect*.

When the sum of all the divisors is three times the number, the number is *triperfect*, or a *perfect number of multiplicity three*.

When the sum of all the divisors is four times the number, the number is a *perfect number of multiplicity four*, and so on.

EXAMPLE

Is 6 a multiply perfect number? If so, of what multiplicity?

$$D_6 = \{1, 2, 3, 6\}$$

$$1 + 2 + 3 + 6 = 12$$

The sum of all the divisors of 6 is 2×6. Thus 6 is a doubly perfect number.

EXERCISE 25

The sets of divisors for four numbers are given below. First find the sum of the divisors of each number. Then determine whether or not the number is multiply perfect, and if so, of what multiplicity.

1. $D_{120} = \{1, 2, 3, 4, 5, 6, 8, 10, 12, 15, 20, 24, 30, 40, 60, 120\}$
2. $D_{200} = \{1, 2, 4, 5, 8, 10, 20, 25, 40, 50, 100, 200\}$
3. $D_{384} = \{1, 2, 3, 4, 6, 8, 12, 16, 24, 32, 48, 64, 96, 128, 192, 384\}$
4. $D_{672} = \{1, 2, 3, 4, 6, 7, 8, 12, 14, 16, 21, 24, 28, 32, 42, 48, 56, 84, 96, 112,$
   $168, 224, 336, 672\}$

Almost Perfect Numbers

An *almost perfect* number is an integer that is 1 more or 1 less than the sum of its proper divisors. An almost perfect number that is 1 more than the sum of its proper divisors is said to be of *type* $+1$. An almost perfect number that is 1 less than the sum of its proper divisors is said to be of *type* -1.

EXAMPLE

Is 4 an almost perfect number? If so, of which type?

$$PD_4 = \{1, 2\} \qquad 1 + 2 = 3$$

The integer 4 is 1 more than the sum of its proper divisors, so 4 is an almost perfect number of type $+1$.

EXERCISE 26

For each number given, compute the sum of its proper divisors. Then determine whether or not the number is almost perfect. If so, state of which type.

1. 8	2. 9	3. 16
4. 32	5. 35	6. 49
7. 55	8. 64	9. 128

Semiperfect Numbers

To be perfect, a number must equal the sum of *all* of its proper divisors. But it is also interesting to examine those numbers that equal the sum of only *some* of their proper divisors.

An integer is *semiperfect* if it is abundant, and if it is possible to find a collection of distinct proper divisors of the integer whose sum equals the integer.

EXAMPLE

Is 12 semiperfect? $PD_{12} = \{1, 2, 3, 4, 6\}$

12 is an abundant number, so we need to find a collection of its proper divisors whose sum is 12.

One choice is $2 + 4 + 6 = 12$. Thus, 12 is semiperfect.

EXERCISE 27

List the proper divisors of each number given. For those numbers that are semi-perfect, show one collection of proper divisors whose sum equals the number.

1. 18	2. 20	3. 24	4. 30	5. 36	6. 40
7. 42	8. 48	9. 56	10. 60	11. 70	

Here is a strategy to apply to find all the ways in which the sum of some proper divisors equals the number.

EXAMPLE

Find all the ways in which the semiperfect number 12 can be written as the sum of a collection of its proper divisors.

$$PD_{12} = \{1, 2, 3, 4, 6\} \quad \text{and} \quad 1 + 2 + 3 + 4 + 6 = 16$$

Since 12 is 4 less than 16, then eliminate 4 from the set of proper divisors to get a sum of 12. This can be done in two ways:

$$1, 2, 3, \cancel{4}, 6 \text{ yields } 1 + 2 + 3 + 6 = 12$$
$$\cancel{1}, 2, \cancel{3}, 4, 6 \text{ yields } 2 + 4 + 6 = 12$$

EXERCISE 28

1. Use the procedure in the example above to find *all* the ways in which each number below can be written as the sum of a collection of its proper divisors.

   a) 18   b) 20   c) 24   d) 30   e) 36

2. Find all the ways in which the sum of a collection of the proper divisors of 72 and 96 equals the number.

Weird Abundant Numbers

A *weird abundant* number is an abundant number that is not semiperfect.

The number 70 in problem 11 of **Exercise 27**, is an example of an abundant number that is not semiperfect. You can verify that 836, 4030, 5830, 7192, 7912, and 9272 are also weird abundant numbers. It has been shown that there are an infinite number of weird abundant numbers.

Operations on Semiperfect Numbers

Now consider what effects the operations of multiplication, addition, and subtraction have on semiperfect numbers.

EXERCISE 29

1. Decide which of the following multiples of 6 are semiperfect.

a) 6	b) 12	c) 18	d) 24	e) 30
f) 66	g) 72	h) 90	i) 120	j) 150
k) 132	l) 216	m) 270	n) 360	o) 420

2. Do you think all multiples of 6 greater than 6 are semiperfect numbers? Explain why or why not.
3. Do you think the number of semiperfect numbers is unlimited? Explain why or why not.
4. Determine if the following multiples of 28 are semiperfect.

   a) $2 \times 28 = 56$ b) $3 \times 28 = 84$
   c) $4 \times 28 = 112$ d) $5 \times 28 = 140$

5. Do you think the product of 3, 4, 5, or more consecutive numbers is semiperfect? Explain why or why not.
6. Compute each of the sums of two semiperfect numbers given below. Then decide whether or not the sum is semiperfect.

a) 30 + 36	b) 12 + 20	c) 42 + 48	d) 20 + 24
e) 40 + 42	f) 72 + 48	g) 80 + 90	h) 120 + 108

7. Is the sum of two semiperfect numbers always semiperfect or just sometimes semiperfect? Explain your answer.
8. Compute each of the differences of two semiperfect numbers given below. Then decide whether or not the difference is semiperfect.

a) 36 − 12	b) 24 − 18	c) 36 − 20	d) 66 − 20
e) 84 − 24	f) 96 − 12	g) 102 − 72	h) 126 − 54

9. Is the difference of two semiperfect numbers always semiperfect or just sometimes semiperfect? Explain your answer.

Primitive Semiperfect Numbers

A *primitive semiperfect number* is a semiperfect number that is not divisible by any other semiperfect number.

Some semiperfect numbers are primitive and some are not. For example, 12 is a primitive semiperfect number because it is not divisible by any other semiperfect number. But 24 is not primitive because 24 is divisible by the semiperfect number 12.

EXERCISE 30

1. Determine whether or not each semiperfect number below is primitive. If it is not primitive, give the semiperfect number that divides it.

   a) 36   b) 56   c) 66   d) 78   e) 88   f) 96
   g) 100   h) 102   i) 138   j) 150   k) 174   l) 180

Amicable Numbers

Is there any connection between the proper divisors of two integers? This question was of special interest to ancient Greek mathematicians. Pythagoras of Samos (6th century, B.C.), made his mark in geometry and number theory. He was the founder of the Pythagorean School whose motto was "all is number." It is quite probable that in about 540 B.C., the Pythagoreans knew a remarkable relationship between the proper divisors of certain pairs of integers.

EXAMPLE

Consider the pair 220 and 284.

$$PD_{220} = \{1, 2, 4, 5, 10, 11, 20, 22, 44, 55, 110\}$$

The sum of the proper divisors of 220 is 284.

$$PD_{284} = \{1, 2, 4, 71, 142\}$$

The sum of the proper divisors of 284 is 220.

Two numbers are *amicable* if each number is the sum of the proper divisors of the other. Amicable numbers are also called *friendly* or *sympathetic* numbers.

In 1980, Elvin J. Lee made an interesting observation about the first amicable number pair (220, 284). He found the sum of the first seventeen primes is $2 \times \underline{220}$ and the sum of the squares of the first seventeen primes is $59 \times \underline{284}$. Can you find some numerical connections to 220 and 284?

EXAMPLE

Are the numbers 1184 and 1210 amicable?

$$PD_{1184} = \{1, 2, 4, 8, 16, 32, 37, 74, 148, 296, 592\}$$

The sum of these proper divisors is 1210.

$$PD_{1210} = \{1, 2, 5, 10, 11, 22, 55, 110, 121, 242, 605\}$$

The sum of these proper divisors is 1184.

So 1184 and 1210 are amicable numbers.

The amicable number pair (1184, 1210) was overlooked for many years by mathematicians. It was finally discovered in 1867 by a 16-year-old Italian boy, B. Nicolo I. Paganini.

The first ten amicable pairs are

(220, 284)	(1184, 1210)
(2620, 2924)	(5020, 5564)
(6232, 6368)	(10744, 10856)
(12285, 14595)	(17296, 18416)
(63020, 76084)	(66928, 66992)

There is no known single formula for deriving all amicable pairs. In 1985, Herman J. J. teRiele of Amsterdam, the Netherlands used an efficient, exhaustive, numerical search algorithm for finding amicable pairs. Using this method all 1427 amicable pairs in which the smaller number is below 10^{10} were identified. By 2007, close to 12 million amicable pairs had been determined.

EXERCISE 31

1. Both numbers in an amicable pair cannot be abundant. Explain why this is so.
2. Can both numbers in an amicable pair be deficient?
   Give a reason for your choice.

Imperfectly Amicable Numbers

> Two numbers are *imperfectly amicable* if the sums of their proper divisors are equal.

The term imperfectly amicable was first used in 1823 by Thomas Taylor (1758–1835) and George Peacock (1791–1858), a professor at Trinity College, Cambridge, England.

EXAMPLE

Are 16 and 33 imperfectly amicable?

$$PD_{16} = \{1, 2, 4, 8\} \quad 1 + 2 + 4 + 8 = 15$$
$$PD_{33} = \{1, 3, 11\} \quad 1 + 3 + 11 = 15$$

Yes, 16 and 33 are imperfectly amicable because the sums of their proper divisors equal 15.

EXERCISE 32

Determine which pairs are imperfectly amicable.

1. (20, 38) 2. (45, 87) 3. (65, 77) 4. (95,119)
5. (69, 133) 6. (93, 145) 7. (115, 187)

Sociable Numbers and Crowds

There are still other patterns based on sums of proper divisors. Suppose a chain of sums of proper divisors is formed in this way. Choose a number to begin, and link it to the sum of its proper divisors. The third number in the chain is the sum of the proper divisors of the second number, and so on. What kinds of chains do perfect and amicable numbers form?

EXAMPLE

Write the chain that begins with 6.

$$\text{Sum } PD_6 = 1 + 2 + 3 = 6.$$

This information is represented in a chain as:

$$6 \xrightarrow{\text{Sum } PD_6} 6$$

The perfect number 6 has a chain with 1 link, or an order-1 chain.

All perfect numbers can be represented by order-1 chains.

EXAMPLE

Write the chain that begins with 220.
Sum $PD_{220} = 284$ Sum $PD_{284} = 220$

This information is represented in a chain as:

$$220 \xrightarrow{\text{Sum } PD_{220}} 284 \xrightarrow{\text{Sum } PD_{284}} 220$$

The amicable numbers 220 and 284 form a chain with 2 links, or an order-2 chain.

All amicable number pairs can be represented by order-2 chains.

> A number is called *sociable* if the chain of sums of proper divisors that leads back to the original number has more than 2 links.

The Belgian Paul Poulet discovered and gave sociable numbers their name in 1918. He wrote about the number with 5 links displayed in the example below. He also provided the number 14,316 that has the largest known number of links, namely 28.

EXAMPLE

Write the chain that starts with 12,496.

$$12,496 - 14,288 - 15,472 - 14,536 - 14,264 - 12,496$$

This chain has 5 links and is an order-5 chain. The number 12,496 is a sociable number.

EXAMPLE

Write the chain that starts with 22.

$$22 - 14 - 10 - 8 - 7 - 1$$

Since 1 has no proper divisors, the chain ends with 1.

The number 22 is not a sociable number because the chain does not return to 22.

Sociable numbers with order-3 chains are called *crowds*.

Are there sociable numbers with order-3 chains? No one has yet found a single crowd, although no one has proved they do not exist.

EXERCISE 33

1. For each number below, write the chain of sums of proper divisors. Indicate the number of links in each chain.

a) 3	b) 9	c) 15	d) 19	e) 20	f) 24	g) 25
h) 27	i) 28	j) 29	k) 30	l) 32	m) 33	n) 34
o) 38	p) 42	q) 45	r) 46	s) 54	t) 60	u) 95

2. How many links are there in a chain that begins with a prime number?

3. Which numbers in problem 1 are sociable numbers?

Practical Numbers

In 1948, A. K. Srinivasan of Mysore, India wrote about a type of numbers he called *practical*. The term "practical" comes from the advantage of using such a number as a base for weights, measures, and coinage. Practical numbers utilize *all* the divisors of a number.

A *practical* number is an integer such that each integer less than or equal to the given integer is a divisor of the given integer or can be written as the sum of distinct divisors of the given integer. Divisors are *not* repeated in the sum. In cases where there are several combinations of divisors that add to the number, the sum with the *least* number of divisors is chosen.

EXAMPLE

Is 12 a practical number? $D_{12} = \{1, 2, 3, 4, 6, 12\}$

Test every integer less than or equal to 12 to see if it is either a divisor of 12 or can be written as a sum of distinct divisors of 12.

1-divisor	2-divisor	3-divisor	4-divisor
$5 = 1 + 4$	6-divisor	$7 = 1 + 6$	$8 = 2 + 6$
$9 = 3 + 6$	$10 = 4 + 6$	$11 = 1 + 4 + 6$	12–divisor

12 is a practical number.

EXAMPLE

Is 8 a practical number? $D_8 = \{1, 2, 4, 8\}$

The integers less than or equal to 8 that are not divisors of 8 can be written as a sum of distinct divisors of 8.

The number 8 is a practical number.

$$3 = 1 + 2 \qquad 5 = 1 + 4 \qquad 6 = 2 + 4 \qquad 7 = 1 + 2 + 4$$

Furthermore, the integers 9 through 15 can also be written as sums of distinct divisors of 8.

$9 = 1 + 8$	$10 = 2 + 8$	$11 = 1 + 2 + 8$	$12 = 4 + 8$
$13 = 1 + 4 + 8$	$14 = 2 + 4 + 8$	$15 = 1 + 2 + 4 + 8$	

Note 15 is the sum of all the divisors of 8.

EXAMPLE

Consider the practical number 12 again.
Can all integers between 12 and 29 be written as sums of distinct divisors of 12?

$$D_{12} = \{1, 2, 3, 4, 6, 12\}$$

Observe

$13 = 1 + 12$	$14 = 2 + 12$	$15 = 3 + 12$
$16 = 4 + 12$	$17 = 1 + 4 + 1$	$18 = 6 + 12$
$19 = 1 + 6 + 12$	$20 = 2 + 6 + 12$	$21 = 1 + 2 + 6 + 12$
$22 = 1 + 3 + 6 + 12$	$23 = 2 + 3 + 6 + 12$	$24 = 2 + 4 + 6 + 12$
$25 = 3 + 4 + 6 + 12$	$26 = 1 + 3 + 4 + 6 + 12$	
$27 = 2 + 3 + 4 + 6 + 12$	$28 = 1 + 2 + 3 + 4 + 6 + 12$	

Note that 28 is the sum of all the divisors of 12.

EXAMPLE

Is 10 a practical number? $D_{10} = \{1, 2, 5, 10\}$

The numbers 4 and 9 cannot be written as a sum of distinct divisors of 10; so 10 is not a practical number.

Try to develop your own method for finding practical numbers.

EXERCISE 34

1. Determine whether or not each of the following powers of 2 is a practical number.

   a) $2^2 = 4$ b) $2^3 = 8$ c) $2^4 = 16$ d) $2^5 = 32$ e) $2^6 = 64$

2. Is the number of practical numbers limited or unlimited? Give a reason for your choice.

3. Determine whether or not each of the following powers of primes is a practical number.

   a) 3^2 b) 3^3 c) 5^2 d) 7^2 e) 7^3
   f) 11^2 g) 11^3 h) 17^2 i) 19^3

4. Does it appear that integer powers of odd primes are practical numbers?

5. Determine whether or not each of the following odd composite numbers is practical.

   a) 15   b) 21   c) 33   d) 35   e) 39   f) 45
   g) 51   h) 55   i) 63   j) 65   k) 69   l) 75

6. Is it possible to have odd practical numbers? Give a reason.

7. Determine whether or not each number below is practical.

   a) 36   b) 38   c) 40   d) 42   e) 46
   f) 50   g) 52   h) 54   i) 56   j) 58
   k) 60   l) 62   m) 64   n) 66   o) 68

8. a) Name 10 abundant numbers that are practical.
   b) Name 10 abundant numbers that are not practical.

EXAMPLE

Is the perfect number 6 a practical number? $D_6 = \{1, 2, 3, 6\}$

The integers 1 through 6 are either divisors of 6 or can be written as sums of divisors of 6. So 6 is a practical as well as a perfect number. Moreover, the sum of the divisors of 6 is 12 and the numbers 7 through 12 can be written as sums of divisors of 6.

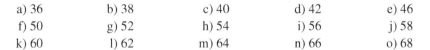

$$7 = 1 + 6 \qquad\qquad 8 = 2 + 6 \qquad\qquad 9 = 3 + 6$$
$$10 = 1 + 3 + 6 \qquad\qquad 11 = 2 + 3 + 6 \qquad\qquad 12 = 1 + 2 + 3 + 6$$

EXERCISE 35

1. List the divisors of 28 and compute their sum.
2. List all the integers from 1 to 28. Where possible, write each integer as a divisor of 28 or as a sum of distinct divisors of 28.
3. Determine if the perfect number 28 is also a practical number.
4. Decide if it is possible to write the integers 29 through 56 as sums of divisors of 28.

Baselike Numbers

> A set of *baselike* numbers for a given integer is a set of numbers such that all numbers less than or equal to the given integer can be written as the sum of some or all of the baselike numbers.

Baselike numbers behave like the divisors of practical numbers. Baselike numbers are also related to the concept of a number base since every number up to a certain point can be written as a sum of baselike numbers.

EXAMPLE

Place 25 one-cent stamps in 5 envelopes, at least one stamp per envelope, so that by selecting one or more of the envelopes and using all of its contents, it will be possible to make stamp value totals for any amount from 1 cent through 25 cents.

A solution to the problem is shown in the following tables.

Envelope number				1	2	3	4	5
Number of one-cent stamps per envelope				1	2	4	8	10

Value in cents	1¢	2¢	3¢	4¢	5¢	...	25¢
Envelope number	1	2	1,2	3	1,3	...	1,2,3,4,5

The numbers 1, 2, 4, 8, and 10 are a set of baselike numbers for the number 25.

EXAMPLE

Buy 5 stamps that total 19 cents so that stamp collection values for any amount from 1 cent through 19 cents can be made.

One possible solution is to buy stamps with values 1¢, 2¢, 4¢, 4¢, and 8¢. This is a set of baselike numbers for 19 in which not all the baselike numbers are different.

Other sets of baselike numbers also solve this problem.

A few of these are (1, 1, 3, 5, 9), (1, 2, 3, 6, 7), and (1, 1, 3, 6, 8).

EXERCISE 36

1. Find three other sets of baselike numbers for the number 25.
2. Place 27 one-cent stamps in 5 envelopes, at least one stamp per envelope, so that by selecting one or more of the envelopes and using all the contents, it will be possible to make stamp value totals for any amount from 1 cent through 27 cents.
3. Buy 6 stamps that total 39 cents so that values for any amount from 1 cent through 39 cents can be made. Find 5 sets of baselike numbers for 39.
4. Buy 7 stamps that total 79 cents so that values for any amount from 1 cent through 79 cents can be made. Find 5 sets of baselike numbers for 79.
5. Buy 8 stamps that total $1.59 so values for any amount from 1 cent through $1.59 can be made. Find 5 sets of baselike numbers for 159.

Chapter 3

Plane Figurate Numbers

I think I have always had a basic liking for the natural numbers. We can conceive of a chemistry which is different from ours, or a biology, but we cannot conceive of a different mathematics of numbers. What is proved about numbers will be a fact in any universe.
Julia Bowman Robinson (1919–1985)

Problems and activities with plane figures come alive when you use manipulatives. Chips, circular discs, or washers on a geoboard are effective tools for modeling as are square and triangular dot paper. The topic of plane figurate numbers is ideal for blending arithmetic, algebra and geometry.

Polygons

> A *polygon* is a closed plane figure formed by three or more line segments called *sides*. Each segment intersects exactly two other segments, one at each endpoint. No two line segments with a common endpoint are on the same line. Each endpoint is called a *vertex* of the polygon.

Each of these figures is a polygon.

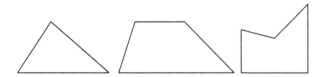

None of these figures is a polygon.

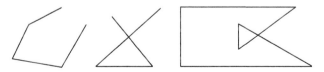

A *regular* polygon is a polygon with all sides congruent and all angles congruent.

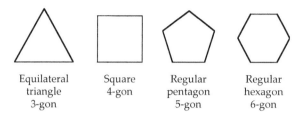

| Equilateral triangle 3-gon | Square 4-gon | Regular pentagon 5-gon | Regular hexagon 6-gon |

EXERCISE 37

1. Use a straightedge or ruler to draw some polygons with 4, 5, and 6 sides that are different from those shown above.
2. What is the angle measure of each angle in a regular
   a) 3-gon?   b) 4-gon?   c) 5-gon?   d) 6-gon?
3. Use a ruler and protractor to draw a regular 8-gon, or octagon.

Figurate Numbers

While the Pythagoreans probably were familiar with both perfect and amicable numbers, they are credited with originating figurate numbers. These numbers are important because they form a link between geometry and arithmetic.

Figurate or *polygonal* numbers are classes of numbers that can be represented by dots arranged in specific geometrical or polygonal patterns.

Triangular Numbers

Triangular numbers are figurate numbers that can be represented by a triangular array of dots.

Triangular numbers

| 1st | 2nd | 3rd | 4th |

```
                                                1   ●
                                  1   ●         2   ●—●
                    1   ●         2   ●—●       3   ●—●—●
        1 ●         2   ●—●       3   ●—●—●     4   ●—●—●—●
        1           1 + 2 = 3     1 + 2 + 3 = 6  1 + 2 + 3 + 4 = 10
```

The numbers 1, 3, 6, 10, 15, 21, ... are triangular numbers. They are the number of dots used in making successive triangular arrays of dots. Note that the triangles used above are equilateral.

One way to find triangular numbers is to add the consecutive integers starting with 1. The total is always a triangular number.

That is,

$1 = 1$, the 1st triangular number,

$1 + 2 = 3$, the 2nd triangular number,

$1 + 2 + 3 = 6$, the 3rd triangular number, and so on.

> A *series* is an indicated sum of a sequence of numbers.
> Each of the numbers in a series is called a *term*.

For example, each of the indicated sums $1 + 2$, $1 + 2 + 3$, and $1 + 2 + 3 + 4$ is a series. The terms are the numbers $1, 2, 3, \ldots$ that are used as addends in the series.

Try the **Investigation: The Super Sum.**

EXERCISE 38

1. Write the series of consecutive integers, starting with 1, that gives each of the numbers below.

   a) 6th triangular number   b) 8th triangular number

   c) 10th triangular number   d) nth triangular number

2. Write the last term in the series of consecutive integers, starting with 1, that gives each of the numbers below.

   a) 35th triangular number   b) 67th triangular number

   c) 100th triangular number   d) nth triangular number

   The sum of a series of consecutive integers, starting with 1, can be found using the method developed in the following examples.

EXAMPLE

Try an odd number of terms in the series:

$$1 + 2 + 3 + 4 + 5 = 15$$

Note that $\qquad\qquad 15 = 5 \quad \times \qquad 3$

$$\uparrow \qquad\qquad\qquad \uparrow$$

number of terms   middle term

in series   in series

Note also that the middle term 3 is the average of the first and last terms,

$$3 = \frac{1 + 5}{2}.$$

The sum equals the number of terms in the series times the middle term in the series. The middle term in a series of consecutive integers with an odd number of terms is the average of the first and last terms.

Here are more illustrations of this method of finding the sum.

$$1 + 2 + 3 + 4 + 5 + 6 + 7 = 7 \times \left(\frac{1 + 7}{2} \right) = 7 \times 4 = 28$$

$$1 + 2 + 3 + \cdots + 25 = 25 \times \left(\frac{1 + 25}{2} \right) = 25 \times 13 = 325$$

A formula for the sum of the series $1 + 2 + \cdots + n$ where n is odd is

$$n\left(\frac{1+n}{2}\right)$$

EXAMPLE

Try an even number of terms in the series.

This presents a slightly different situation because there is no middle term in a series with an even number of terms.

$$1 + 2 + 3 + 4 = 2 \times (1 + 4) = 2 \times 5 = 10$$

$$\uparrow \qquad \uparrow$$

number   sum of

of pairs   each pair

$$1 + 2 + 3 + 4 + 5 + 6 = 3 \times (1 + 6) = 3 \times 7 = 21$$

$$\uparrow \qquad \uparrow$$

number   sum of

of pairs   each pair

Here the numbers are paired beginning with the pair of the first and last terms. Notice that each pair has the same sum, and that the number of pairs is half the number of terms in the series, or half the last term.

The number of pairs in $1+2+3+4$ is $\dfrac{4}{2} = 2$, and the sum of each pair is $1+4 = 5$.

The number of pairs in $1 + 2 + 3 + 4 + 5 + 6$ is $\dfrac{6}{2} = 3$, and the sum of each pair is $1 + 6 = 7$.

A formula for the sum of the series $1 + 2 + \cdots + n$ where n is even is

$$\frac{n}{2}(1 + n)$$

After examining the two formulas developed in the examples above, it is clear a *single* formula results for the sum of any series of consecutive integers starting with one.

$$1 + 2 + 3 + \cdots + n = n\left(\frac{1+n}{2}\right)$$

$$= \frac{n}{2}(n + 1)$$

$$= \frac{n(n + 1)}{2}$$

There is a compelling anecdote drawn from the life story of a young boy who became a mathematical giant. Carl Friedrich Gauss (1777–1855) was born in Brunswick, Germany. Young Carl attended a one-room school and the tale is told that one day his teacher, in order to rest a bit, asked the pupils to find the sum of the consecutive integers from 1 to 100. Unfortunately for the teacher, Carl completed the task rapidly and brought forward his slate showing only the answer, which was correct. Since that day there has been considerable speculation on how Gauss calculated the total. Gauss had an illustrious career in mathematics and earned the distinction of being "one of the greatest mathematicians of all time."

Since the nth triangular number can be written as the sum of the integers from 1 through n, there is a formula to calculate the value of any triangular number.

$$T_n = n\text{th triangular number} = \frac{n(n+1)}{2}$$

EXERCISE 39

1. Find the first 30 triangular numbers.

2. Is the number of triangular numbers finite or infinite? Explain why.

Operations on Triangular Numbers

Are the sum, difference, and product of triangular numbers also triangular numbers? Let's try some examples.

EXAMPLE

Are the sum, difference and product of the triangular numbers 3 and 6 also triangular?

The sum, $3 + 6 = 9$, is not triangular.

The difference, $6 - 3 = 3$, is triangular.

The product, $3 \times 6 = 18$, is not triangular.

There are many directions to take with triangular numbers, some of which are explored in the following examples. For instance, can *any* integer be written as a triangular number or as a sum of triangular numbers? If repetitions of triangular numbers in the sums arc permitted, what is the least number of triangular numbers needed to express the sums?

EXAMPLE

Write the first ten integers as the sums of the *least* number of triangular numbers.

$1 = 1$ $2 = 1 + 1$ $3 = 3$ $4 = 1 + 3$ $5 = 1 + 1 + 3$
$6 = 6$ $7 = 1 + 6$ $8 = 1 + 1 + 6$ $9 = 3 + 6$ $10 = 10$

Can each triangular number greater than 1 be written as a sum of triangular numbers if repetitions are allowed?

EXAMPLE

Write the 2nd through the 7th triangular number as sums of triangular numbers using the *least* number of terms.

$3 = 1 + 1 + 1$ $6 = 3 + 3$ $10 = 1 + 3 + 6$
$15 = 3 + 6 + 6$ $21 = 6 + 15$ $28 = 1 + 6 + 21$

EXERCISE 40

1. a) Is the sum of two triangular numbers always, sometimes, or never triangular?

    b) Is the difference of two triangular numbers always, sometimes, or never triangular?

    c) Is the product of two triangular numbers always, sometimes, or never triangular?

2. Write each integer below as a sum of the *least* number of triangular numbers.

    a) 11   b) 14   c) 34   d) 37   e) 40   f) 43
    g) 51   h) 68   i) 75   j) 82   k) 94   l) 100

3. In problem 2, what is the *maximum* number of triangular numbers used in any sum?

4. Using two or more triangular numbers, write each of the triangular numbers below as a sum of the *least* number of triangular numbers.

    a) 36   b) 55   c) 78   d) 91   e) 120
    f) 153   g) 190   h) 231   i) 300

5. In problem 4, what is the maximum number of triangular numbers used in any sum?

EXAMPLE

Find four triangular numbers that can be written as sums of *exactly four* triangular numbers.

$$6 = 1 + 1 + 1 + 3 \qquad\qquad 10 = 1 + 3 + 3 + 3$$
$$15 = 3 + 3 + 3 + 6 \qquad\qquad 21 = 3 + 6 + 6 + 6$$

EXERCISE 41

Try to write each triangular number below as a sum of *exactly four* triangular numbers.

1. 28   2. 45   3. 66
4. 105   5. 136   6. 171
7. 210   8. 276   9. 325

Perfect Numbers, Triangular Numbers, and Sums of Cubes

The numbers 6, 28, and 496 are perfect numbers. They are also triangular numbers.

$$1 + 2 + 3 = 6$$
$$1 + 2 + \cdots + 7 = 28$$
$$1 + 2 + \cdots + 31 = 496$$

It is a fact that all even perfect numbers are also triangular numbers.

There is an interesting pattern for the squares of the first three triangular numbers.

Triangular number	Square of triangular number	Relation between square of triangular number and sum of cubes of integers
1	$1^2 = 1$	$1^3 = 1 = 1^2 = (1)^2$
3	$3^2 = 9$	$1^3 + 2^3 = 9 = 3^2 = (1+2)^2$
6	$6^2 = 36$	$1^3 + 2^3 + 3^3 = 6^2 = (1+2+3)^2$

EXERCISE 42

1. Find the square of each triangular number below. Try to write the square as the sum of the cubes of consecutive integers beginning with 1.

   a) 10   b) 15   c) 21   d) 28   e) 36   f) 45   g) 55

2. **Challenge** Use the formula for triangular numbers,
$$T_n = n\frac{(n+1)}{2}$$
Show that all even perfect numbers are triangular numbers.

Pascal's Triangle

The Pascal triangle was not created by the French mathematician, Blaise Pascal (1623–1662), for whom it is named. The existence of the mathematics in the triangle and then the triangle itself can be traced back over centuries beginning with the 2nd century B.C. to various individuals from India, China and Persia. Pascal's recognition comes from his writings printed in 1654 and which were published posthumously in 1665. His publication was called the **Traité du triangle arithmétique**.

Pascal's triangle, a special triangular array of integers, is rich in number patterns. The triangle can be formed quite simply as shown.

Row	Pascal's triangle
0	1
1	1 1
2	1 2 1
3	1 3 3 1
4	1 4 6 4 1

In this array

a) the first and last number in each row is always 1;
b) beginning with row 2 and excepting the 1's, each number in a row is the sum of the two numbers in the row above as marked.

EXERCISE 43

1. Copy Pascal's triangle and give the entries in rows 5, 6, 7, 8, 9, and 10.
2. Study the diagonals in Pascal's triangle. Do you see any patterns? Can you identify the types of numbers in any of the diagonals? Explain.
3. Represent Pascal's triangle as a right triangle as shown below.

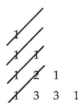

Continue the triangle in this form through row 10.
Draw seven more rising diagonals so that each diagonal

a) begins with 1 and
b) is parallel to the other diagonals.

Sum the numbers on each diagonal. Do you notice any patterns in the totals? Explain.

Try the **Investigation: Conjecturing with Pascal.**

Triangle Inequality Numbers

Triangle inequality numbers come from a fundamental fact about triangles, and present several challenging problems.

An *equilateral* triangle is a triangle in which all three sides are congruent. An *isosceles* triangle is a triangle in which two sides are congruent. A *scalene* triangle is a triangle in which no two sides are congruent.

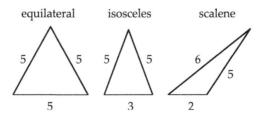

equilateral   isosceles   scalene

What three line segments can form a triangle? A little experimentation leads to this result.

Three line segments can form the sides of a triangle only if the sum of the lengths of any two sides is greater than the length of the third side.

In the scalene triangle shown above, $2 + 6 > 5$, $2 + 5 > 6$, and $5 + 6 > 2$.

Triangle inequality numbers are three numbers such that the corresponding lengths form a triangle.

For example, $(1, 1, 1)$ are equilateral triangle inequality numbers, $(3, 3, 4)$ are isosceles triangle inequality numbers, and $(3, 5, 6)$ are scalene triangle inequality numbers.

EXERCISE 44

1. List four equilateral triangle inequality numbers.
2. List four isosceles triangle inequality numbers.
3. List four scalene triangle inequality numbers.

EXAMPLE

Given a set of lengths corresponding to $1, 2, 3$, how many different sets of triangle inequality numbers can be made?

There are 3 equilateral triangle inequality numbers:

$$(1, 1, 1), \quad (2, 2, 2), \quad (3, 3, 3)$$

There are 4 isosceles triangle inequality numbers:

$$(2, 2, 1), \quad (2, 2, 3), \quad (3, 3, 1), \quad (3, 3, 2)$$

There are 0 scalene triangle inequality numbers.

EXAMPLE

Given a set of lengths corresponding to $1, 2, 3, 4$, how many different sets of triangle inequality numbers can be made?

There are 4 equilateral triangle inequality numbers:

$$(1, 1, 1), \quad (2, 2, 2), \quad (3, 3, 3), \quad (4, 4, 4).$$

There are 8 isosceles triangle inequality numbers:

$$(2, 2, 1), \quad (2, 2, 3), \quad (3, 3, 1), \quad (3, 3, 2),$$
$$(3, 3, 4), \quad (4, 4, 1), \quad (4, 4, 2), \quad (4, 4, 3).$$

There is 1 set of scalene triangle inequality numbers:

$$(2, 3, 4).$$

The information in the two preceding examples can be summarized in the following table.

Given lengths corresponding to numbers	Number of equilateral triangles	Number of isosceles triangles	Number of scalene triangles	Total number of triangles
1, 2, 3	3	4	0	7
1, 2, 3, 4	4	8	1	13

EXERCISE 45

1. Make a table like the one above and record information about sets of given lengths from (1, 2, 3, 4, 5) through (1, 2, 3, 4, 5, 6, 7, 8, 9, 10).
2. **Challenge** Study the table in problem 1 and look for patterns. By examining the numbers in the columns can you extend the table a few more rows? Verify your choices. Can you suggest a rule or formula to predict the number of equilateral triangles for a given set of numbers? How about a rule for isosceles triangles? Scalene triangles?
3. Let T_n be the nth triangular number and set $T_0 = 0$. The triangular numbers can be labeled as shown.

T_0	T_1	T_2	T_3	T_4	T_5	T_6	T_7	T_8	T_9	T_{10}	T_{11}	T_{12}
0	1	3	6	10	15	21	28	36	45	55	66	78

Make a table like the one that follows. Use the format of the examples done for you. Complete the table for sets of given lengths from (1, 2, 3, ... , 9) through (1, 2, 3, ... , 14).

Given lengths corresponding to numbers	Number of scalene triangle inequality numbers is given by
1, 2, 3	$T_0 = 0$
1, 2, 3, 4	$T_1 = 1$
1, 2, 3, ... , 5	$T_2 = 3$
1, 2, 3, ... , 6	$T_1 + T_3 = 1 + 6 = 7$
1, 2, 3, ... , 7	$T_2 + T_4 = 3 + 10 = 13$
1, 2, 3, ... , 8	$T_1 + T_3 + T_5 = 1 + 6 + 15 = 22$

Given lengths corresponding to numbers	Number of scalene triangle inequality numbers is given by
1, 2, 3, . . . , 9	
1, 2, 3, . . . , 10	
1, 2, 3, . . . , 11	
1, 2, 3, . . . , 12	
1, 2, 3, . . . , 13	
1, 2, 3, . . . , 14	

4. Make a table like the one used in problem 1. Record information about the following lengths that correspond to *random* sets of consecutive numbers.

a) 10, 11, 12
c) 8, 9, 10, 11, 12
e) 21, 22, 23, 24, 25, 26, 27

b) 5, 6, 7, 8
d) 15, 16, 17, 18, 19, 20
f) 45, 46, 47, 48, 49, 50, 51, 52

Rectangular Numbers

Single integers were called *linear* numbers by the early Greeks. Plato (429–348 B.C.) mentioned two kinds of linear number products, the square numbers and the rectangular or oblong numbers. The latter were called *plane* numbers and this wording was used by Euclid about 300 B.C. Rectangular numbers are figurate numbers but they do not belong to the regular polygonal class of figurate numbers.

> *Rectangular* numbers are figurate numbers that can be represented by a rectangular array of dots.

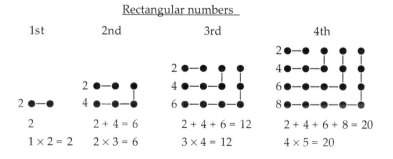

The numbers 2, 6, 12, 20, . . . are rectangular numbers. They are the number of dots used in making successive rectangular arrays of dots in which the number of rows and columns increase by one. There are various ways to find rectangular numbers.

1. *Rectangular numbers are sums of consecutive even numbers beginning with 2.*

   $2 = 2$ 1st rectangular number
   $2 + 4 = 6$ 2nd rectangular number
   $2 + 4 + 6 = 12$ 3rd rectangular number
   $2 + 4 + 6 + 8 = 20$ 4th rectangular number

   Note that the sum for the 3rd rectangular number begins with 2 and has 3 terms, and the sum for the 4th rectangular number begins with 2 and has 4 terms. Following this pattern, the sum for the nth rectangular number begins with 2 and has n terms that are consecutive even numbers.

2. *Rectangular numbers are twice the corresponding triangular numbers.*
   The sum pattern for the 1st rectangular number through the 4th rectangular number can be written as shown.

   1st rectangular number: $2 = 2(1) = 2$
   2nd rectangular number: $2 + 4 = 2(1 + 2) = 2(3) = 6$
   3rd rectangular number: $2 + 4 + 6 = 2(1 + 2 + 3) = 2(6) = 12$
   4th rectangular number: $2 + 4 + 6 + 8 = 2(1 + 2 + 3 + 4) = 2(10) = 20$

   It is clear that the first four rectangular numbers are twice the corresponding triangular numbers. Check this pattern with other rectangular numbers.

   This pattern, the nth rectangular number equals 2 times the nth triangular number, is evident in the figure below. In each case, the rectangular array of dots has been split into two equal triangular arrays.

   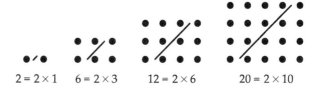

   $2 = 2 \times 1$ $6 = 2 \times 3$ $12 = 2 \times 6$ $20 = 2 \times 10$

3. *Rectangular numbers are the product of two consecutive numbers.*

   1st rectangular number: $2 = 1 \times 2$
   2nd rectangular number: $6 = 2 \times 3$
   3rd rectangular number: $12 = 3 \times 4$
   4th rectangular number: $20 = 4 \times 5$

   The nth rectangular number is the product of the two consecutive numbers n and $n + 1$; that is, the nth rectangular number equals $n(n + 1)$.

   The Greeks also recognized as rectangular numbers those numbers that are products of two numbers in the form $n(n + m)$ with $n = 1, 2, 3, \ldots$, and

$m = 2, 3, 4, \ldots$. Thus, for $m = 2$, the rectangular numbers are $3, 8, 15,$ $24, 35, \ldots$, and for $m = 3$, the rectangular numbers are $4, 10, 18, 28, 40, \ldots$.

EXERCISE 46

1. Write each of the rectangular numbers below in two ways: as a sum of consecutive even numbers, and as a product of two consecutive numbers. Compute the value of each rectangular number.

   a) 5th rectangular number   b) 6th rectangular number
   c) 7th rectangular number   d) 8th rectangular number
   e) 9th rectangular number   f) 10th rectangular number
   g) 12th rectangular number   h) 15th rectangular number
   i) 18th rectangular number   j) 20th rectangular number

2. The values of the triangular numbers T_5 through T_{16} are given below. Compute the value of the 5th rectangular number through the 16th rectangular number using the pattern: nth rectangular number $= 2 \times n$th triangular number

   a) $T_5 = 15$ b) $T_6 = 21$ c) $T_7 = 28$ d) $T_8 = 36$
   e) $T_9 = 45$ f) $T_{10} = 55$ g) $T_{11} = 66$ h) $T_{12} = 78$
   i) $T_{13} = 91$ j) $T_{14} = 105$ k) $T_{15} = 120$ l) $T_{16} = 136$

3. Two plane rectangular numbers $a \times b$ and $c \times d$ are similar plane rectangular numbers if their sides are proportional, that is if $a : b = c : d$, or $a/b = c/d$, or $a \times d = b \times c$. The plane rectangular numbers $6 = 2 \times 3$ and $24 = 4 \times 6$ are similar because $2 \times 6 = 3 \times 4$, or $2 : 3 = 4 : 6$. Create five sets of similar plane rectangular numbers.

Square Numbers

Square numbers are figurate numbers that belong to the regular polygon class of figurate numbers.

> *Square* numbers are figurate numbers that can be represented by a square array of dots.

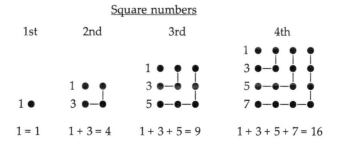

The numbers $1, 4, 9, 16, \ldots$ are square numbers. The nth square number is represented by an n by n array of dots. So the value of the nth square number is $n \times n$ or n^2. The successive square arrays of dots are also produced by adding consecutive odd numbers of dots. Starting with the first square number 1, the 2nd square number is the sum of the odd numbers $1 + 3$, the 3rd square number is the sum of the odd numbers $1 + 3 + 5$, and so on.

EXAMPLE

Circle the odd numbers and look for patterns.

$$\textcircled{1}, 2, \textcircled{3}, 4, \textcircled{5}, 6, \textcircled{7}, 8, \textcircled{9}$$

The 3rd odd number is 5; it is given by $5 = 6 - 1 = (2 \times 3) - 1$.
The 4th odd number is 7; it is given by $7 = 8 - 1 = (2 \times 4) - 1$.
The 5th odd number is 9; is given by $9 = 10 - 1 = (2 \times 5) - 1$.
The pattern for the value of the nth odd number is $(2 \times n) - 1$ or $2n - 1$.

EXAMPLE

Find a pattern for computing the sum of a series of consecutive odd numbers that begins with 1.
Compute $1 + 3 + 5 + \cdots + (2n - 1)$.

This pattern can be developed in a way similar to the one for the sum of a series of consecutive integers that begins with 1.

- Suppose that the series has an odd number of terms. $1 + 3 + 5 + 7 + 9 = 25$

$$\text{Note that } 25 = 5 \ \times \ 5$$
$$\uparrow \qquad \uparrow$$
$$\text{number of} \qquad \text{middle}$$
$$\text{terms in series} \qquad \text{term}$$

The middle term is the average of the first and last terms, and the sum is the product of the number of terms and the middle term. So for odd values of n, the sum is $n \left(\dfrac{1 + 2n - 1}{2} \right)$ which simplifies to n^2.

- Suppose the series has an even number of terms. $1 + 3 + 5 + 7 = 16$

$$\text{Note that } 16 = 2 \ \times \ 8$$
$$\uparrow \qquad \uparrow$$
$$\text{number of} \qquad \text{sum of}$$
$$\text{pairs in series} \qquad \text{each pair}$$

The number of pairs in a series with n terms where n is even is $\dfrac{n}{2}$. The sum of each pair is equal to the sum of the first term, 1, and the last term, $2n - 1$. So for even values of n, the sum of the series is the number of pairs times the sum of each pair, or $\dfrac{n}{2}(1 + 2n - 1)$ which again simplifies to n^2.

Thus, $1 + 3 + 5 + \cdots + (2n - 1) = n^2$.

EXERCISE 47

1. Compute the value of each of the square numbers below using the pattern:

$$n\text{th square number} = n^2$$

    a) 14th square number   b) 18th square number

    c) 21st square number   d) 25th square number

2. Decide whether the number of square numbers is limited or unlimited. Give a reason.

3. Write each square number below as a series of consecutive odd numbers beginning with 1. Indicate the number of terms in each series.

    a) 2nd square number   b) 3rd square number

    c) 6th square number   d) 10th square number

    e) 12th square number   f) 20th square number

4. For each of the following series of consecutive odd numbers, indicate the number of terms and the sum of the terms in the series. Determine which square number is represented by each series.

    a) $1 + 3 + \cdots + 13$ b) $1 + 3 + \cdots + 15$

    c) $1 + 3 + \cdots + 17$ d) $1 + 3 + \cdots + 23$

    e) $1 + 3 + \cdots + 25$ f) $1 + 3 + \cdots + 31$

5. Study the figures below.

$1 = 1$ $1 + 3 = 4$ $3 + 6 = 9$ $6 + 10 = 16$

The numbers 1, 3, 6, and 10 are triangular numbers. It appears that, excepting the first, each square number is the sum of two consecutive triangular numbers. In the case of the 4th square number, it is the sum of the 3rd and 4th triangular numbers.

Write each square number below as the sum of two consecutive triangular numbers.

    a) 5th square number   b) 6th square number

    c) 7th square number   d) 8th square number

    e) 9th square number   f) 10th square number

    g) 15th square number   h) nth square number

Sums of Square Numbers

Integers can be written as sums of at most three triangular numbers. Can any integer be written as a square number or as a sum of square numbers? Suppose that repetitions of square numbers in the sum are permitted. What is the *least* number of square numbers needed to express the sums?

EXAMPLE

Write each of the first six integers as a sum of the *least* number of square numbers.

$$1 = 1^2 \qquad\qquad 2 = 1^2 + 1^2 \qquad\qquad 3 = 1^2 + 1^2 + 1^2$$
$$4 = 2^2 \qquad\qquad 5 = 1^2 + 2^2 \qquad\qquad 6 = 1^2 + 1^2 + 2^2$$

EXAMPLE

Explore the relationship between the sums of two consecutive squares and the triangular numbers.

$$1^2 + 2^2 = 1 + 4 = 5 \qquad \text{and} \qquad 5 = (1)4 + 1 = (1)4 + 1$$
$$2^2 + 3^2 = 4 + 9 = 13 \qquad \text{and} \qquad 13 = (1 + 2)4 + 1 = (3)4 + 1$$
$$3^2 + 4^2 = 9 + 16 = 25 \qquad \text{and} \qquad 25 = (1 + 2 + 3)4 + 1 = (6)4 + 1$$

EXERCISE 48

1. Write each of the integers from 6 through 19 as a sum of the *least* number of square numbers.
2. a) In problem 1, what is the *maximum* number of squares used in any sum?
   b) Since triangular numbers are integers, each triangular number can be written as a sum of square numbers. What is the *maximum* number of squares in any such sum?
3. The example above illustrates a pattern between the sum of two consecutive square numbers and the triangular numbers. Check to see if the pattern continues for the sums that follow.

   a) $4^2 + 5^2$ b) $5^2 + 6^2$ c) $6^2 + 7^2$
   d) $7^2 + 8^2$ e) $8^2 + 9^2$ f) $9^2 + 10^2$
   g) $10^2 + 11^2$ h) $14^2 + 15^2$ i) $19^2 + 20^2$

4. In a sentence generalize the pattern connecting consecutive square numbers and triangular numbers that you found in problem 3.

Try the **Investigations: Footsteps of Lagrange** and **A Medieval Pattern**.

Positive Square Pair Numbers

Squares are the focus of many interesting problems. The following examples describe a connection between square numbers and sets of consecutive integers that begin with 1.

EXAMPLE

Let $A = \{1, 2, 3\}$ and $B = \{1, 2, 3\}$. Sets A and B contain the same numbers 1, 2, 3. Is it possible to pair the numbers in A with those in B so that the sum of each pair is a square number?

elements from A:	1	2	3
elements from B:	$+3$	$+2$	$+1$
	4	4	4

Since 4 is a square number, then for this choice of the sets A and B, a pairing is possible.

We say that 3, the largest number in each set, is called a *positive square pair number*.

> A number n is a *positive square pair number* if it is possible to pair off the numbers from 1 to n with the numbers 1 to n so that the sum of each pair is a square.

EXAMPLE

Is 4 a positive square pair number?

Consider the sets $A = \{1, 2, 3, 4\}$ and $B = \{1, 2, 3, 4\}$. It is not possible to pair the numbers in A with those in B so that the sum of each pair is a square number.

Hence, 4 is not a positive square pair number.

EXAMPLE

Is 5 a positive square pair number?
Consider the sets $A = B = \{1, 2, 3, 4, 5\}$.

elements from A:	1	2	3	4	5
elements from B:	$+3$	$+2$	$+1$	$+5$	$+4$
	4	4	4	9	9

Since 4 and 9 are square numbers, 5 is a positive square pair number.

EXERCISE 49

1. Determine whether each of the following numbers is a positive square pair number. Show the pairings of numbers in sets A and B.

   a) 1; $A = B = \{1\}$

   b) 2; $A = B = \{1, 2\}$

   c) 6; $A = B = \{1, 2, \ldots, 6\}$

   d) 7; $A = B = \{1, 2, \ldots, 7\}$

e) 8; $A = B = \{1, 2, \ldots, 8\}$ f) 9; $A = B = \{1, 2, \ldots, 9\}$
g) 10; $A = B = \{1, 2, \ldots, 10\}$

2. Which of the numbers less than 11 are positive square pair numbers?
3. **Challenge** How many other positive square pair numbers can you find?

Bigrade Numbers

Bigrade numbers are pairs of number triples that satisfy two requirements.

1. The sums of the numbers in each triple are equal.
2. The sums of the squares of the numbers in each triple are equal.

The name *bigrade* is also applied to quadruples, quintuples, and their extensions.

EXAMPLE

Are the triples $(1, 5, 6)$ and $(2, 3, 7)$ bigrade numbers?

First, compute the sums of the numbers in the two triples. $1 + 5 + 6 = 12$ and $2 + 3 + 7 = 12$.

Second, compute the sums of the squares of the numbers in the two triples. $1^2 + 5^2 + 6^2 = 62$ and $2^2 + 3^2 + 7^2 = 62$

The triples $(1, 5, 6)$ and $(2, 3, 7)$ are bigrade numbers.

EXERCISE 50

1. Which of the following pairs of triples are bigrade numbers?

   a) $(1, 6, 9)$ and $(1, 7, 8)$ b) $(2, 4, 9)$ and $(1, 6, 8)$
   c) $(3, 4, 5)$ and $(3, 3, 6)$ d) $(2, 2, 5)$ and $(1, 4, 4)$

2. Which of the two pairs of triples are bigrade numbers?

   a) $(2134, 4284, 6374)$ and $(2154, 4244, 6394)$
   b) $(3421, 8442, 7463)$ and $(5421, 4442, 9463)$

3. Are the quintuples $(1, 15, 45, 153, 171)$ and $(3, 6, 66, 120, 190)$ bigrade numbers?
4. **Challenge** Find some other bigrade triples.

Pythagorean Triples

Pythagorean triples are triples of integers such that the sum of the squares of the two smaller integers equals the square of the third integer.

EXAMPLE

Is the triple (3, 4, 5) a Pythagorean triple?
$$3^2 + 4^2 = 25 \quad \text{and} \quad 5^2 = 25$$
(3, 4, 5) is a Pythagorean triple.

EXAMPLE

Is the triple (5, 12, 13) a Pythagorean triple?
$$5^2 + 12^2 = 169 \quad \text{and} \quad 13^2 = 169$$
(5, 12, 13) is a Pythagorean triple.

This application of square numbers has a long history. Pythagorean triples are named after the Greek mathematician, Pythagoras. But the triples were put to practical use even before the time of Pythagoras. In ancient Egypt, surveyors, sometimes called rope stretchers, would take a rope that had equally-spaced knots in it and tie the rope around three stakes as shown below. The distance from stake 1 to stake 2 is 3 units, from stake 2 to stake 3 is 4 units, and from stake 3 to stake 1 is 5 units. By doing this, the surveyors knew they had constructed a right triangle in which the right angle was between the two sides of lengths 3 and 4 units.

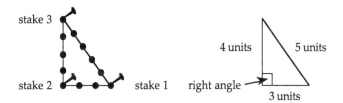

It can be shown that if a right triangle has integer sides, then the three integers will be a Pythagorean triple. It is also true that if the sides of a triangle are a Pythagorean triple, then the triangle will be a right triangle.

There is some historical evidence that in addition to the Greeks and Egyptians, the ancient Babylonians knew how to generate triples of integers as in the examples above. It is interesting to speculate just how to go about finding these Pythagorean triples.

EXERCISE 51

Determine if each of the following triples of numbers is a Pythagorean triple. Problems 1 and 2 were completed for you.

1. (3, 4, 5)   2. (5, 12, 13)
3. (7, 24, 25)   4. (8, 15, 17)
5. (20, 21, 29)   6. (9, 40, 41)
7. (12, 35, 37)   8. (28, 45, 53)
9. (33, 56, 65)   10. (36, 77, 85)

11. Find out more about Pythagoras and his school. Write a report on your findings.

Primitive Pythagorean Triples

There are two kinds of Pythagorean triples. Read the example below carefully.

EXAMPLE

Show that the 2-, 3-, and 4-multiples of the Pythagorean triple (3, 4, 5) are also Pythagorean triples.
Consider the 2-multiples of (3, 4, 5).

$2(3) = 6$, $2(4) = 8$, $2(5) = 10$
$6^2 + 8^2 = 10^2$ (6, 8, 10) is a Pythagorean triple.

Consider the 3-multiples of (3, 4, 5).

$3(3) = 9$, $3(4) = 12$, $3(5) = 15$
$9^2 + 12^2 = 15^2$ (9, 12, 15) is a Pythagorean triple.

Consider the 4-multiples of (3, 4, 5).

$4(3) = 12$, $4(4) = 16$, $4(5) = 20$
$12^2 + 16^2 = 20^2$ (12, 16, 20) is a Pythagorean triple.

The only common factor of the numbers in the Pythagorean triple (3, 4, 5) is 1. Numbers in multiples of Pythagorean triples have other common factors. The numbers in the Pythagorean triple (12, 16, 20) in the example above have common factors 1, 2, and 4.

> *Primitive* Pythagorean triples are Pythagorean triples in which the three numbers have only a common factor of 1.
> *Nonprimitive* Pythagorean triples are Pythagorean triples in which the three numbers have common factors in addition to 1.

EXERCISE 52

1. List the nonprimitive Pythagorean triples from **EXERCISE 51**.
2. Show that the 2- and 3-multiples of (8, 15, 17) are Pythagorean triples.
3. Show that the 10- and 20-multiples of (5, 12, 13) are Pythagorean triples.
4. Show that any multiple of a Pythagorean triple is also a Pythagorean triple.
5. Look for patterns in the primitive Pythagorean triples listed below. Then answer each question about primitive Pythagorean triples.

   (5, 12, 13)   (7, 24, 25)   (8, 15, 17)   (20, 21, 19)
   (12, 35, 37)   (28, 45, 53)   (33, 56, 65)

   a) Is the largest number in a triple odd?
   b) Can the two smaller numbers in a triple both be even?
   c) Can all three numbers in a triple be odd?

d) Is one of the two smaller numbers in a triple divisible by 3?

e) Is one of the two smaller numbers in a triple divisible by 4?

6. Find five primitive Pythagorean triples different from those in problem 5 and in **EXERCISE 51**.

7. **Challenge** Sums of two consecutive squares that equal a square number can be found among Pythagorean triples. For instance, there is (3, 4, 5) and (20, 21, 29). Can you find others like these? Can you find three consecutive squares whose sum is a square number? Four consecutive squares? Five consecutive squares?

Several formulas have been developed that produce primitive Pythagorean triples (a, b, c). Two such formulas are described in the **Investigation: Pythagorean Triple Pursuits**.

Congruent Numbers

Pythagorean triples can be used to find *congruent* numbers.

EXAMPLE

Is there a number n such that $5^2 + n$ and $5^2 - n$ are both square numbers?

1. Find a Pythagorean triple in which 5 is the largest number. Here we take (3, 4, 5).
2. Form the product $2(3 \times 4) = 24$
3. Compute $5^2 + 24 = 49$, a square number.
   Compute $5^2 - 24 = 1$, a square number.
4. n is 24 and is called a *congruent* number.

EXAMPLE

Is there a number n such that $13^2 + n$ and $13^2 - n$ are both square numbers?

1. Find a Pythagorean triple in which 13 is the largest number. Here we take (5, 12, 13).
2. Form the product $2(5 \times 12) = 120$.
3. Compute $13^2 + 120 = 289$, a square number.
   Compute $13^2 - 120 = 49$, a square number.
4. n is 120 and is a *congruent* number.

> A number that is twice the product of the two smaller numbers in a Pythagorean triple is called a *congruent* number.

Every Pythagorean triple produces a congruent number.

Do you think the number of congruent numbers is unlimited?

EXERCISE 53

1. Find the congruent number produced by each of the Pythagorean triples (a, b, c). Verify in each case that $c^2 + n$ and $c^2 - n$ are square numbers.

    a) (6, 8, 10)   b) (8, 15, 17)   c) (12, 16, 20)
    d) (7, 24, 25)   e) (20, 21, 29)   f) (10, 24, 26)

2. Is there a number n such that
    a) $15^2 + n$ and $15^2 - n$ are both square numbers? If so, name n.
    b) $30^2 + n$ and $30^2 - n$ are both square numbers? If so, name n.

Fermat's Last Theorem

We learned Pythagorean triples are three integers such that the sum of the *squares* of two of the integers equals the square of the third integer. Around 1637, the French mathematician, Pierre de Fermat, stated,

"No triple of integers can be found such that the sum of the cubes, the 4th powers, the 5th powers, . . . , of two of the integers is equal in turn to the cube, the 4th power, the 5th power, . . . , of the third integer."

Fermat meant there were no triples of integers (a, b, c) such that $a^n + b^n = c^n$ for $n = 3, 4, 5, \ldots$. In the margin of the book, **Arithmetica** by Diophantus, Fermat wrote a statement in Latin which included the famous remark . . .

"I have assuredly found an admirable proof of this, but the margin is too narrow to contain it."

Fermat died leaving no record of the proof and no one else was able to prove or disprove Fermat's Last Theorem until the 1990s. In 1993, Andrew Wiles of Princeton University announced a 200-page proof to the international mathematical community. But it was found to contain a flaw. Additional work was carried out by Wiles and his former student, Richard Taylor, of Cambridge University, and in 1995, the paper proving Fermat's Last Theorem was published.

Happy Numbers

Happy numbers are an application of square numbers.

> A *happy* number is an integer for which the sum of the squares of the digits eventually ends in 1.

EXAMPLE

Is 13 a happy number?
Compute the sum of the squares of the digits of 13.

$$1^2 + 3^2 = 1 + 9 = 10$$

Stop if the result is 1; otherwise, repeat the process.

Compute the sum of the squares of the digits of 10.

$$1^2 + 0^2 = 1 + 0 = 1$$

Stop, since the result is 1.

13 is a happy number. Since 13 is a happy number, it follows that 31 is also a happy number. Why?

EXAMPLE

Is 4 a happy number?

$$4^2 = 16$$
$$1^2 + 6^2 = 1 + 36 = 37$$
$$3^2 + 7^2 = 9 + 49 = 58$$
$$5^2 + 8^2 = 25 + 64 = 89$$
$$8^2 + 9^2 = 64 + 81 = 145$$
$$1^2 + 4^2 + 5^2 = 1 + 16 + 25 = 42$$
$$4^2 + 2^2 = 16 + 4 = 20$$
$$2^2 + 0^2 = 4 + 0 = 4$$

Since the starting number 4 reappeared the pattern above will repeat over and over, and will never end in 1. Thus 4 is not a happy number.

EXERCISE 54

1. Find all happy numbers less than 100. Be sure to find and use a short cut strategy.

Operations on Happy Numbers

As you can see in the following examples, sums, differences, and products of happy numbers are only sometimes happy numbers themselves.

EXAMPLE

Consider the sums of happy numbers.

In these cases, the sum of two happy numbers *is* a happy number:

$$10 + 13 = 23 \qquad 1 + 31 = 32 \qquad 13 + 31 = 44 \qquad 19 + 49 = 68$$

In these cases, the sum of two happy numbers *is not* a happy number:

$$7 + 10 = 17 \qquad 1 + 13 = 14 \qquad 10 + 19 = 29 \qquad 7 + 23 = 30$$

EXAMPLE

Consider the differences of happy numbers.

In these cases, the difference of two happy numbers *is* a happy number:

$$32 - 1 = 31 \qquad 44 - 31 = 13 \qquad 68 - 49 = 19 \qquad 23 - 13 = 10$$

In these cases, the difference of two happy numbers *is not* a happy number:

$$28 - 13 = 15 \qquad 32 - 10 = 22 \qquad 79 - 68 = 11 \qquad 82 - 44 = 38$$

EXAMPLE

Consider the products of happy numbers.

In these cases, the product of two happy numbers *is* a happy number:

$$7 \times 10 = 70 \qquad 7 \times 13 = 91 \qquad 10 \times 13 = 130 \qquad 10 \times 28 = 280$$

In these cases, the product of two happy numbers *is not* a happy number:

$$7 \times 23 = 161 \qquad 13 \times 19 = 247 \qquad 13 \times 23 = 299 \qquad 19 \times 23 = 437$$

EXERCISE 55

1. Find a pair of happy numbers different from the ones in the examples such that

   a) The sum is a happy number
   b) The sum is not a happy number
   c) The difference is a happy number
   d) The difference is not a happy number
   e) The product is a happy number
   f) The product is not a happy number

2. **Challenge** Can you find any patterns in the sums, differences, and products of happy numbers that yield happy numbers?

Happy Number Words

Assign the numbers 1, 2, 3, ... , 26 to the letters of the English alphabet as shown in the chart.

A	B	C	D	E	F	G	H	I	J	K	L	M
1	2	3	4	5	6	7	8	9	10	11	12	13

N	O	P	Q	R	S	T	U	V	W	X	Y	Z
14	15	16	17	18	19	20	21	22	23	24	25	26

EXAMPLE

Is TINA a happy number name?

Sum the numbers matched with each letter of the name.

$$T \quad I \quad N \quad A$$
$$20 + 9 + 14 + 1 = 44$$

Since 44 *is* a happy number, TINA is a happy number name.

Is JANUARY a happy number month?

$$J \quad A \quad N \quad U \quad A \quad R \quad Y$$
$$10 + 1 + 14 + 21 + 1 + 18 + 25 = 90$$

Since 90 *is not* a happy number, JANUARY is not a happy number month.

EXAMPLE

Is 2003 a happy number year?

$$2^2 + 3^2 = 4 + 9 = 13, \text{ a happy number.}$$

Since 13 *is* a happy number, 2003 is a happy number year.

EXERCISE 56

1. Which of the following are happy number names?

   a) BRIAN   b) KATYA   c) JIA   d) CHIN
   e) KEN   f) HOPE   g) TAJ   h) LUIS
   i) MARIA   j) GINA   k) RINA   l) HUAN

2. Check to see if your first name, middle name, last name, and complete name are happy number names. Try the names of some friends.
3. Which months of the year are happy number months?
4. Which days of the week are happy number days?
5. Which of the following years are happy number years?

   a) 1981   b) 1999   c) 2000   d) 2008
   e) 2030   f) 2105   g) 2111   h) 2190

Repeating Cycles

Many numbers are not happy numbers. These numbers, like 4, go through a repeating cycle. Remember that 4 goes through the repeating cycle of eight numbers: 4, 16, 37, 58, 89, 145, 42, 20. Other numbers that are not happy will also enter the above repeating cycle at some point.

EXAMPLE

Is 29 a happy number?

$$2^2 + 9^2 = 4 + 81 = 85$$
$$8^2 + 5^2 = 64 + 25 = 89$$

29 is not a happy number and enters the repeating cycle at cycle number 89. That is, 29 goes through the pattern 29, 85, <u>89</u>, 145, 42, 20, 4, 16, 37, 58, <u>89</u>, and so on.

EXAMPLE

Is 33 a happy number?

$$3^2 + 3^2 = 9 + 9 = 18$$
$$1^2 + 8^2 = 1 + 64 = 65$$
$$6^2 + 5^2 = 36 + 25 = 61$$
$$6^2 + 1^2 = 36 + 1 = 37$$

33 is not a happy number and enters the repeating cycle at cycle number 37. That is, 33 goes through the pattern 33, 18, 65, 61, <u>37</u>, 58, 89, 145, 42, 20, 4, 16, <u>37</u>, and so on.

EXERCISE 57

1. Find the cycle number at which each of the following numbers enters the repeating cycle.

   a) 24   b) 39   c) 51   d) 78   e) 73
   f) 38   g) 25   h) 17   i) 77

2. Find other numbers that enter the repeating cycle at each of the eight cycle numbers.

Patterns in Squares of 1, 11, 111, ...

An interesting pattern involving the middle digit appears in the squares of numbers whose digits are all 1s.

$$1^2 = 1$$
$$11^2 = 121$$
$$111^2 = 12321$$

EXERCISE 58

1. In a sentence, describe the pattern you see in the middle digit of the squares above.

2. Find each square number below.

a) $1,111^2$ b) $11,111^2$ c) $111,111^2$
d) $1,111,111^2$ e) $11,111,111^2$ f) $111,111,111^2$

3. Does the pattern in problem 1 hold for each number in problem 2? Will the pattern always hold? Explain.

Squarefree Numbers

A *squarefree* number is an integer that is not divisible by the square of any prime. Alternatively, an integer is a squarefree number if 1 is the only square that divides it.

The first six squarefree numbers are 1, 2, 3, 5, 6, and 7. The numbers 4, 8, 9, 12, and 16 are not squarefree because 4, 8, 12, and 16 are all divisible by 2^2 and 9 is divisible by 3^2.

How can you decide whether an integer is squarefree if you have expressed it as a product of primes?

EXAMPLE

Write the integers 2, 3, 4, 5, and 20 as the sum of two squarefree numbers.

$$2 = 1 + 1 \quad 3 = 1 + 2 \quad 4 = 1 + 3 = 2 + 2 \quad 5 = 2 + 3$$

The number 20 can be written as the sum of two squarefree numbers in several ways.

$$20 = 1 + 19 = 3 + 17 = 5 + 15 = 6 + 14 = 7 + 13 = 10 + 10$$

EXAMPLE

Write the integers 1, 2, 4, and 8 as a product of a square number and a squarefree number.

$$1 = 1^2 \times 1 \quad 2 = 1^2 \times 2 \quad 4 = 2^2 \times 1 \quad 8 = 2^2 \times 2$$

EXERCISE 59

1. Write each number below as the sum of two squarefree numbers.

a) 6   b) 7   c) 8   d) 9   e) 10
f) 11   g) 32   h) 43   i) 54   j) 65

2. Write each of the following numbers as the sum of two squarefree numbers in *all* possible ways.

a) 21   b) 22   c) 23   d) 24   e) 25

3. Write each number below as the product of a square number and a squarefree number.

a) 10   b) 11   c) 12   d) 13   e) 14
f) 42   g) 60   h) 175   i) 210   j) 294

Tetragonal Numbers

A *tetragon* is a nonsquare plane figure having four angles and four sides.

Tetragonal numbers are figurate numbers but they do not belong to the regular polygonal class of figurate numbers.

A *tetragonal* number is a figurate number that can be represented by the particular array of dots shown below.

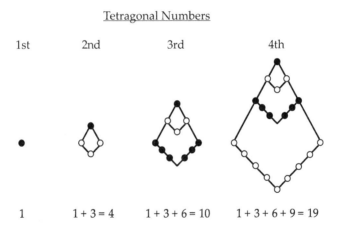

Tetragonal Numbers

1st   2nd   3rd   4th

1   1 + 3 = 4   1 + 3 + 6 = 10   1 + 3 + 6 + 9 = 19

The numbers 1, 4, 10, 19, ... are tetragonal numbers. They are the number of dots used in making successive tetragonal arrays in which the number of dots equals the sum of 1 and consecutive multiples of 3. Thus, a way to find tetragonal numbers is to start with 1 and add consecutive multiples of 3.

EXERCISE 60

1. Write each tetragonal number from the 5th through the 10th as a series. Compute the sum of each series.
2. It appears from the table on p. 101 that the tetragonal numbers are sums of triangular and square numbers.

Extend the table through the 9th tetragonal number.

Triangular number	Square number	Tetragonal number
1st : 1	—	1st: 1=1
2nd: 3	1st : 1	2nd: $3 + 1 = 4$
3rd: 6	2nd: 4	3rd: $6 + 4 = 6 + 3 + 1 = 10$

3. Tell why the number of tetragonal numbers is limited or unlimited.
4. List the units digit of each of the first 25 tetragonal numbers. Which digits do not appear in your list? Is there any pattern? Discuss.
5. Develop a formula that gives the tetragonal numbers.

Pentagonal Numbers

Pentagonal numbers are figurate numbers that can be represented by dots arranged in the form of a regular pentagon.

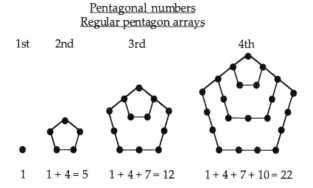

Pentagonal numbers
Regular pentagon arrays

1st	2nd	3rd	4th
1	$1 + 4 = 5$	$1 + 4 + 7 = 12$	$1 + 4 + 7 + 10 = 22$

The numbers $1, 5, 12, 22, \ldots$ are pentagonal numbers. They are the number of dots used in making successive pentagonal arrays of dots. In each case the number of pentagons increases by one, and the number of dots on each side of the largest pentagon increases by one.

Study the pattern in the series that gives the pentagonal numbers.

1st pentagonal number: 1 = 1

2nd pentagonal number: 1 + 4 = 5
                                    3
3rd pentagonal number: 1 + 4 + 7 = 12
                                        3
4th pentagonal number: 1 + 4 + 7 + 10 = 22
                                            3
5th pentagonal number: 1 + 4 + 7 + 10 + 13 = 35
                                                3

What series represents the *n*th pentagonal number?

Here is another way to view the pattern.

1st pentagonal number: 1 = 1

2nd pentagonal number: (1) + (1 + 3) = 5

3rd pentagonal number: (1 + 4) + (4 + 3) = 12

4th pentagonal number: (1 + 4 + 7) + (7 + 3) = 22

5th pentagonal number: (1 + 4 + 7 + 10) + (10 + 3) = 35

What series represents the *n*th pentagonal number here?

Every integer, including the pentagonal numbers, can be written as the sum of at most four square numbers. Also every integer, including the pentagonal numbers, can be written as the sum of at most three triangular numbers. For example, the pentagonal number 35 can be written as the sum of the three square numbers $1 + 9 + 25$ and as the sum of the three triangular numbers $1 + 6 + 28$.

EXERCISE 61

1. Write each of the following pentagonal numbers as a sum of the *least* number of squares.

   a) 51   b) 70   c) 92   d) 117   e) 145
   f) 176   g) 210   h) 247   i) 287

2. Name the smallest pentagonal number that is the sum of *exactly four* square numbers.

3. Write each of the following pentagonal numbers as a sum of the *least* number of triangular numbers.

   a) 51   b) 70   c) 92   d) 117   e) 145
   f) 176   g) 210   h) 247   i) 287

4. Can you name a pentagonal number that is the sum of *exactly two* triangular numbers?

5. Write each of the following pentagonal numbers as the sum of *three* triangular numbers, *two* of which are the same.

a) 22   b) 35   c) 51   d) 70   e) 92
f) 117   g) 145   h) 176   i) 210   j) 247

6. Describe a pattern you notice in the results of problem 5.

Hexagonal Numbers

Hexagonal numbers are figurate numbers that can be represented by dots arranged in the form of a regular hexagon.

Hexagonal numbers
Regular hexagonal arrays

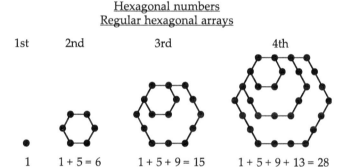

1st   2nd   3rd   4th

1   1 + 5 = 6   1 + 5 + 9 = 15   1 + 5 + 9 + 13 = 28

The numbers 1, 6, 15, 28, ... are hexagonal numbers. They are the number of dots used in making successive hexagonal arrays of dots. In each case the number of hexagons increases by one and the number of dots on each side of the largest hexagon increases by one.

Study the pattern in the series that gives the hexagonal numbers.

1st hexagonal number: $1 = 1$

2nd hexagonal number: $\underset{4}{1 + 5} = 6$

3rd hexagonal number: $1 + \underset{4}{5 + 9} = 15$

4th hexagonal number: $1 + 5 + \underset{4}{9 + 13} = 28$

5th hexagonal number: $1 + 5 + 9 + \underset{4}{13 + 17} = 45$

Here is another way to view the pattern.

1st hexagonal number: $1 = 1$

2nd hexagonal number: $(1) + (1 + 4) = 6$

3rd hexagonal number: $(1 + 5) + (5 + 4) = 15$

4th hexagonal number: $(1 + 5 + 9) + (9 + 4) = 28$

5th hexagonal number: $(1 + 5 + 9 + 13) + (13 + 4) = 45$

Generalize the pattern for the nth hexagonal number in each case.

A different way to consider the hexagonal numbers is to rearrange the dots into a nonregular hexagon array.

Hexagonal numbers
Nonregular hexagon array

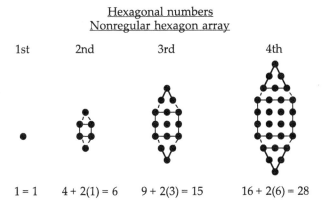

1st   2nd   3rd   4th

$1 = 1$ $4 + 2(1) = 6$ $9 + 2(3) = 15$ $16 + 2(6) = 28$

When represented as a nonregular hexagon array of dots, each hexagonal number (allowing 0 to be a triangular number) is the sum of a square number and twice a triangular number.

If the hexagonal numbers are represented by a nonregular hexagon array and grouped as shown below, we have yet another way of looking at them.

Hexagonal numbers
Nonregular hexagon array

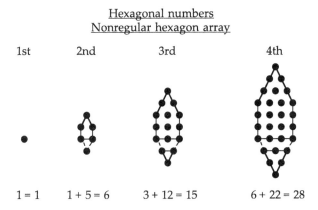

1st   2nd   3rd   4th

$1 = 1$ $1 + 5 = 6$ $3 + 12 = 15$ $6 + 22 = 28$

In this arrangement of the nonregular hexagon array, each hexagonal number (allowing 0 to be a triangular number) is the sum of a triangular number and a pentagonal number.

EXERCISE 62

1. The 5th hexagonal number can be written as the series $1 + 5 + 9 + 13 + 17$. Write each hexagonal number from the 6th through the 16th as a series. Find the sum of each series.
2. Draw a *nonregular* hexagonal array for the 5th and 6th hexagonal numbers.
   a) Does the pattern of square and triangular numbers hold in these arrays?
   b) Does the pattern of triangular and pentagonal numbers hold?
3. In the table below, each hexagonal number is written as the sum of a square and twice a triangular number. Extend the table through the 16th hexagonal number.

Square Number	Triangular number	Hexagonal number
1st: 1	—	1st: $1 = 1$
2nd: 4	1st: 1	2nd: $4 + 2(1) = 6$
3rd: 9	2nd: 3	3rd: $9 + 2(3) = 15$
4th: 16	3rd: 6	4th: $16 + 2(6) = 28$

4. In the table below, each hexagonal number is written as the sum of a triangular number and a pentagonal number. Extend the table through the 16th hexagonal number.

Triangular number	Pentagonal number	Hexagonal number
—	1st: 1	1st: $1 = 1$
1st: 1	2nd: 5	2nd: $1 + 5 = 6$
2nd: 3	3rd: 12	3rd: $3 + 12 = 15$
3rd: 6	4th: 22	4th: $6 + 22 = 28$

5. In the table on p. 106, it appears that the product of an odd number and an integer is a hexagonal number. Extend the table through the 12th row of entries. Check to see if you get a hexagonal number in each case.

Odd number	Integer	Hexagonal number
1st: 1	1st: 1	1st: $1 \times 1 = 1$
2nd: 3	2nd: 2	2nd: $3 \times 2 = 6$
3rd: 5	3rd: 3	3rd: $5 \times 3 = 15$

6. The first 12 triangular numbers are:

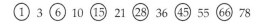

   ① 3 ⑥ 10 ⑮ 21 ㉘ 36 ㊺ 55 ⑥⑥ 78

   Are the circled numbers hexagonal numbers? Do you think that this pattern of
   locating hexagonal numbers within the list of triangular numbers will continue?
   Give a reason.

7. Decide if the number of hexagonal numbers is limited or unlimited. Give a
   reason.

8. List the units digit of each of the first 25 hexagonal numbers. Which digits do
   not appear on your list? Is there any pattern? Discuss.

9. Find a formula that gives the hexagonal numbers.

Every integer, including the hexagonal numbers, can be written as the sum of at
most four square numbers. Also every integer, including the hexagonal numbers,
can be written as the sum of at most three triangular numbers. For example, the
hexagonal number 45 can be written as the sum of two squares $9 + 36$ and as the
sum of three triangular numbers $3 + 6 + 36$.

EXERCISE 63

1. Write each of the following hexagonal numbers as a sum of the *least* number
   of squares.

   a) 66   b) 91   c) 120   d) 153   e) 190   f) 231
   g) 276   h) 325   i) 378   j) 435   k) 496

2. Name the smallest hexagonal number that is the sum of *exactly* four square
   numbers.

3. Write each of the following hexagonal numbers as a sum of the *least* number
   of triangular numbers. Note: a single number is *not* considered a sum in this
   problem.

   a) 66   b) 91   c) 120   d) 153   e) 190   f) 231
   g) 276   h) 325   i) 378   j) 435   k) 496

4. The first three hexagonal numbers can be represented by the nonregular hexagon
   array of dots shown in figure (a). This hexagonal pattern can be rearranged to

form the triangular pattern in figure (b). Represent the 4th, 5th, and 6th hexagonal numbers in this triangular pattern, if possible. What does this action imply?

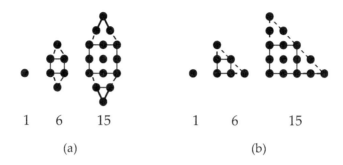

1 6 15	1 6 15
(a)	(b)

5. The numbers 6, 28, and 496 are even perfect numbers and they are also hexagonal numbers. Can you show that every even perfect number is also a hexagonal number?

Recursion and Figurate Numbers

Plane figurate number types are related to one another in a variety of ways. For example, all figurate number types can be expressed as sums of triangular numbers. Examine specific figurate numbers for other connections. Each plane figurate number type has a generalization that can be given in *recursive* form and in *explicit* or *closed* form.

> The nth term of a sequence is in *recursive* form if each new member of the sequence is expressed in terms of the preceding member(s). The nth term of a sequence is in *explicit* or *closed* form when it is expressed in terms of n.

EXAMPLE

Represent the triangular numbers in recursive form and in closed form.
Since

$$T_1 = 1$$
$$T_2 = 1 + 2 = T_1 + 2$$
$$T_3 = (1 + 2) + 3 = T_2 + 3$$

then the recursive form for T_n appears to be

$$T_n = T_{n-1} + n$$

The closed form for triangular numbers developed earlier is

$$T_n = \frac{n(n+1)}{2}$$

EXAMPLE

Write an explicit form expression for pentagonal numbers using the fact that pentagonal numbers are sums of triangular numbers.

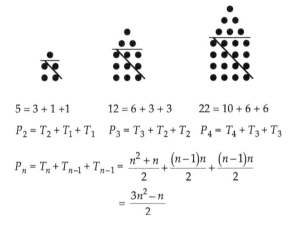

$$5 = 3 + 1 + 1 \qquad 12 = 6 + 3 + 3 \qquad 22 = 10 + 6 + 6$$

$$P_2 = T_2 + T_1 + T_1 \quad P_3 = T_3 + T_2 + T_2 \quad P_4 = T_4 + T_3 + T_3$$

$$P_n = T_n + T_{n-1} + T_{n-1} = \frac{n^2 + n}{2} + \frac{(n-1)n}{2} + \frac{(n-1)n}{2}$$

$$= \frac{3n^2 - n}{2}$$

EXERCISE 64

1. Fill in the first five terms and give the recursive form representation for each figurate number type.

# of Sides	Name	First Five Terms					Recursive Formula
3	Triangular	1	3	6	10	15	$T_n = T_{n-1} + n$
4	Rectangular	2	6	12	20	30	$R_n = R_{n-1} + 2n$
4	Square	1	4	9	16	25	$S_n = S_{n-1} + 2n - 1$
5	Pentagonal	1	5				
6	Hexagonal	1	6				
7	Heptagonal	1	7				
8	Octagonal	1	8				
9	Nonagonal	1	9				
10	Decagonal	1	10				
11	Undecagonal	1	11				
12	Dodecagonal	1	12				
13	Tridecagonal	1	13				
14	Tetradecagonal	1	14				
15	Pentadecagonal	1	15				
16	Hexadecagonal	1	16				
17	Heptadecagonal	1	17				
18	Octadecagonal	1	18				
19	Nondecagonal	1	19				
20	Icosagonal	1	20				

2. Using the fact that figurate numbers are sums of triangular numbers, find a closed form representation for each figurate number type in the table in problem 1.

Remainder Patterns in Figurate Numbers

Consider only the figurate numbers associated with four regular polygons: the equilateral triangle, the square, the regular pentagon, and the regular hexagon. In each case the corresponding set of numbers is divided term by term by 3, 4, 5, and 6 respectively.

The remainders are recorded in the table below. Do you notice any patterns in the table? If so, can you describe them?

Numbers				Remainders after division			
Tri.	Square	Pent.	Hex.	Tri. $\div$ 3	Square $\div$ 4	Pent. $\div$ 5	Hex. $\div$ 6
1st: 1	1	1	1	1	1	1	1
2nd: 3	4	5	6	0	0	0	0
3rd: 6	9	12	15	0	1	2	3

EXERCISE 65

1. Extend the above table through the 12th row of entries.
2. Use the data in problem 1 to answer the following questions.

   a) What is the repetition pattern of the remainders of the triangular numbers after division by 3? How many numbers are in the repetition pattern?
   b) What is the repetition pattern of the remainders of the square numbers after division by 4? How many numbers are in the repetition pattern?
   c) What is the repetition pattern of the remainders of the pentagonal numbers after division by 5? How many numbers are in the pattern?
   d) What is the repetition pattern of the remainders of the hexagonal numbers after division by 6? How many numbers are in the pattern?

Geometric dot patterns can be drawn for many other figurate or polygonal numbers. The triangular, rectangular, square, tetragonal, pentagonal, and hexagonal numbers and their properties give some basic ideas about the class of figurate numbers. Continue to study and make new discoveries about figurate numbers.

Gnomic Numbers

Sequences of numbers in which the terms differ from each other by a constant are called *arithmetic sequences* or *arithmetic progressions*. For example,

$1, 2, 3, 4, 5, \ldots$ with first term 1 and constant difference 1, and $1, 3, 5, 7, 9, \ldots$ with first term 1 and constant difference 2 are arithmetic sequences.

In general, the terms of an arithmetic progression are

$$a, \ a + d, \ a + 2d, \ a + 3d, \ a + 4d, \ldots$$

where a is the first term and d is the common difference.

Each term of an arithmetic progression is called a *gnomic* number or a *gnomon*.

EXAMPLE

Show that the integers are gnomic numbers for the triangular numbers.

The triangular numbers are sums of the consecutive terms in the arithmetic sequence $1, 2, 3, 4, \ldots$.

Notice that adding 2 and 3 dots to the preceding array does not change the shape of the figure. The integers are gnomic numbers for the triangular numbers.

EXAMPLE

Show that the odd numbers are gnomic numbers for the square numbers.

The square numbers are sums of the consecutive odd numbers $1, 3, 5, 7, \ldots$.

Notice that adding 3 and 5 dots to the preceding array does not change the shape of the figure. The odd numbers are gnomic numbers for the square numbers.

EXERCISE 66

1. Make an arithmetic sequence that

   a) starts with 1 and has a common difference of 3.
   b) starts with 1 and has a common difference of 6.
   c) starts with 1 and has a common difference of 10.
   d) starts with 5 and has a common difference of 4.
   e) starts with 4 and has a common difference of 5.

2. a) Give the first six gnomic numbers for the pentagonal numbers.
   b) Give the first six gnomic numbers for the hexagonal numbers.

Lo-Shu (Luoshu) Magic Square; Male and Female Numbers

Apparently the Chinese originated the representation of numbers in simple geometric patterns long before the time of Pythagoras of Samos. Popular myth claims that the legendary Emperor Yu saw a magic square on the back of a tortoise along the banks of the Luo River during a great flood around 2000 B.C. Study the array on the tortoise shell below. White circle patterns represented the odd numbers and black circle patterns represented the even numbers. The magic square used the numbers 1 through 9 arranged so that the numbers in the three rows and three columns and two diagonals had the same sum, 15. This sum is called the magic constant.

Magic square

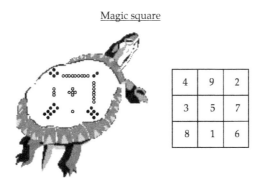

Observe that the four even numbers are positioned in the four corners of the square, while the five odd numbers appear in the middle row and column of the square. Notice also that 5 is in the center cell of the square and that three times this middle number equals the magic constant. Check to see that using rotations and reflections of the square there are exactly eight ways to represent the numbers 1 through 9 in a magic square.

In the days of the Greek philosopher and mathematician, Pythagoras, the number 1 was considered to be the source of all numbers. The other odd numbers were *male* numbers while the even numbers were *female* numbers.

Try these **Investigations** related to content developed in this chapter: **Hexagons in Black and White, Pentagonal Play, Centered Triangular Numbers, Triangular Number Turnarounds,** and **Trying Trapezoids.**

Chapter 4

Solid Figurate Numbers

Number is the ruler of forms and ideas and the cause of gods and demons.

Pythagoras (6th Century B.C.)

The study of solid figurate numbers as well as plane figurate numbers provides a setting that integrates geometry with arithmetic and algebra.

Polyhedra and Solid Figurate Numbers

The set of solid figurate numbers requires some geometric background and terms. A *geometric solid* is a closed surface enclosing a portion of space.

Geometric solids

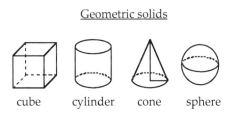

cube   cylinder   cone   sphere

If the surfaces of a geometric solid are polygonal, the solid is called a *polyhedron*. That is, a geometric solid composed of four or more polygonal surfaces, no two of which are in the same plane, is a polyhedron. The polygonal surfaces are the *faces* of the polyhedron; the intersections of the faces are the *edges* of the polyhedron; the intersections of the edges are the *vertices* of the polyhedron.

A *right circular cylinder* (shown above) is a solid with the bases circular and with the axis joining the two centers of the bases perpendicular to the planes of the two bases. A *right circular cone* is a solid (as above) bounded by a circular base in a plane and by a surface joining a vertex, the apex, to the edge of the base. A *sphere* (above) is a solid such that all points on its surface are equidistant from a fixed point called the *center.*

Polyhedra are often called *solid figures* or simply *solids*.

<u>Polyhedra</u>

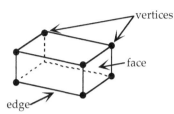

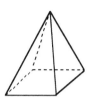

Polyhedra are classified and named by the number of faces.

Number of faces	Name of polyhedron
4	tetrahedron
5	pentahedron
6	hexahedron
7	heptahedron
8	octahedron
9	nonahedron
10	decahedron
11	undecahedron
12	dodecahedron
20	icosahedron

A polyhedron is said to be *regular* if its faces are congruent, regular polygons and if all of its polyhedral angles are congruent. There are exactly five regular polyhedra. These have been known since the time of the Greek philosopher Plato (429–348 B.C.) and are often called *Platonic Solids*.

<u>Regular polyhedra</u>

tetrahedron   hexahedron or cube   octahedron

dodecahedron   icosahedron

Observe that of these regular polyhedra, the tetrahedron has 4 *triangular* faces, the hexahedron has 6 *square* faces, the octahedron has 8 *triangular* faces, the dodecahedron has 12 *pentagonal* faces, and the icosahedron has 20 *triangular* faces.

The names *space* numbers, *spatial* numbers, and *solid figurate* numbers all refer to the same set of numbers. Solid figurate numbers are the counterpart of plane figurate numbers in space. Several kinds of solid figurate numbers will be investigated in the following sections.

Pyramidal Numbers

The grounds in front of an armory or city hall are sometimes decorated with cannon balls piled in the form of a pyramid. In markets, fruit with spherical shape such as oranges, grapefruit, and apples are often displayed in pyramidal stacks. The pyramid is a practical way of piling these objects.

Several types of pyramidal pilings give rise to different types of pyramidal numbers.

Tetrahedral Numbers, Triangular Pyramidal Numbers

Tetrahedral (or triangular pyramidal) numbers are the simplest kind of pyramidal numbers. To visualize these numbers it will help to imagine stacks of oranges, balls or other sphere-like shapes. The base and layers of a tetrahedral piling are equilateral triangular in shape.

EXAMPLE

In a tetrahedral piling with four layers, layer 1 has one sphere and it rests on three spheres in layer 2. The three spheres in layer 2 rest on six spheres in layer 3, which rest on ten spheres in layer 4.

Layer	Number of Spheres per Layer	
1	1	
2	3	
3	6	
4	10	

Tetrahedral numbers are the sums of consecutive triangular numbers.

The first four tetrahedral or triangular pyramidal numbers are 1, 4, 10, and 20.

Tetrahedral number	Number of layers in piling	Total number of spheres
1st	1	1
2nd	2	$1 + 3 = 4$
3rd	3	$1 + 3 + 6 = 10$
4th	4	$1 + 3 + 6 + 10 = 20$

Each new tetrahedral number can be expressed in terms of the one preceding it. Let TH_k be the kth tetrahedral number and recall that T_k is the kth triangular number.

Tetrahedral number	Number of layers in piling	Total number of spheres
TH_1	1	1
TH_2	2	$TH_1 + T_2 = 1 + 3$
TH_3	3	$TH_2 + T_3 = 1 + 3 + 6$
TH_4	4	$TH_3 + T_4 = 1 + 3 + 6 + 10$

Since the number of triangular numbers is unlimited, the number of tetrahedral numbers is also unlimited.

Tetrahedral numbers can also be represented by dots appropriately arranged on a tetrahedron. The fourth tetrahedral number is pictured below.

It is possible to describe tetrahedral numbers using the quotient rule illustrated in the example.

EXAMPLE

1st tetrahedral number: $\dfrac{1(2)(3)}{6} = 1$

2nd tetrahedral number: $\dfrac{2(3)(4)}{6} = 4$

3rd tetrahedral number: $\dfrac{3(4)(5)}{6} = 10$

4th tetrahedral number: $\dfrac{4(5)(6)}{6} = 20$

Notice that in the numerator of each quotient, the k in TH_k is the first term of the product.

EXERCISE 67

1. Look for patterns in the table below. Then extend the table through the 12th tetrahedral number.

   Locate the tetrahedral numbers in the Pascal triangle.

Tetrahedral number	Value
1st	$1 = 1$
2nd	$1 + 3 = 4$
3rd	$4 + 6 = 10$
4th	$10 + 10 = 20$

2. Use the quotient rule to compute the values of each of the tetrahedral numbers below.

   a) 5th   b) 6th   c) 7th   d) 8th   e) 10th
   f) 15th   g) 18th   h) 20th   i) 25th   j) 50th

3. Determine a formula that gives the tetrahedral numbers.

Square Pyramidal Numbers

To visualize square pyramidal numbers, it will help to imagine stacks of spheres. The base and layers of a square pyramidal piling are square in shape.

EXAMPLE

In a square pyramidal piling with four layers, layer 1 has one sphere and it rests on four spheres in layer 2. The four spheres in layer 2 rest on nine spheres in layer 3, which rest on 16 spheres in layer 4.

Layer	Number of Spheres per Layer	
1	1	
2	4	
3	9	
4	16	

Square pyramidal numbers are the sums of consecutive square numbers.

Square pyramidal number	Number of layers in piling	Total number of spheres
1st	1	1
2nd	2	$1 + 4 = 5$
3rd	3	$1 + 4 + 9 = 14$
4th	4	$1 + 4 + 9 + 16 = 30$

Each square pyramidal number can be expressed in terms of the one preceding it. The first four square pyramidal numbers are 1, 5, 14, and 30. Since there are an unlimited number of square numbers, there are an unlimited number of square pyramidal numbers.

Square pyramidal numbers can also be represented by dots appropriately arranged on a square pyramid. The fourth square pyramidal number is pictured below. Notice that the fourth square pyramidal number can be expressed as the sum of the third and fourth tetrahedral numbers.

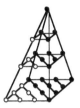

EXERCISE 68

1. Look for patterns in the table below. Then extend the table through the 10th square pyramidal number.

Square pyramidal numbers	Value
1st	1
2nd	$1 + 4 = 5$
3rd	$5 + 9 = 14$
4th	$14 + 16 = 30$
5th	$30 + 25 = 55$

2. Make a drawing showing that the fifth square pyramidal number is the sum of the fourth and fifth tetrahedral numbers.
3. Make tetrahedral and square pyramidal number models using translucent marbles and glue.
   Try the **Investigation: Marble Art.**
4. Look for patterns in the table. Then extend the table through the 9th square pyramidal number.

Square Pyramidal numbers	Value as the sum of tetrahedral numbers
2nd	$\dfrac{1(2)(3)}{6} + \dfrac{2(3)(4)}{6}$
3rd	$\dfrac{2(3)(4)}{6} + \dfrac{3(4)(5)}{6}$

5. Determine a formula that gives the square pyramidal numbers.

Pentagonal Pyramidal Numbers

Tetrahedral and square pyramidal numbers can be illustrated by uniform symmetrical pyramids of symmetrical objects such as cannonballs, oranges, and marbles. In fact, such symmetrical pyramids are possible only for pyramids with triangular or square bases. The base of a pentagonal pyramidal piling is pentagonal but this does not produce a uniform symmetrical pyramid.

Pentagonal pyramidal numbers are the sums of consecutive pyramidal numbers.

Since the number of pentagonal numbers is unlimited, the number of pentagonal pyramidal numbers is also unlimited.

EXERCISE 69

1. Look for patterns in the table below. Then extend the table through the 12th pentagonal pyramidal number.

Pentagonal pyramidal number	Value
1st	1
2nd	$1 + 5 = 6$
3rd	$6 + 12 = 18$
4th	$18 + 22 = 40$

2. Determine a formula that gives the pentagonal pyramidal numbers.

Hexagonal Pyramidal Numbers

The base of a hexagonal pyramidal piling is hexagonal and, like the pentagonal pyramid, does not result in a uniform symmetrical pyramid.

Hexagonal pyramidal numbers are the sums of consecutive hexagonal numbers.

Since the number of hexagonal numbers is unlimited, the number of hexagonal pyramidal numbers is also unlimited.

EXERCISE 70

1. Look for patterns in the table on p. 121. Then extend the table through the 12th hexagonal pyramidal number.

Hexagonal pyramidal number	Value
1st	1
2nd	$1 + 6 = 7$
3rd	$7 + 15 = 22$
4th	$22 + 28 = 50$

2. Determine a formula that gives the hexagonal pyramidal numbers.

Heptagonal and Octagonal Pyramidal Numbers

The base of the heptagonal pyramidal piling is regular heptagonal in shape and the base of the octagonal pyramidal piling is regular octagonal in shape. The same pattern holds for all other pyramidal pilings that involve regular polygons.

Heptagonal pyramidal numbers are the sums of consecutive heptagonal numbers. *Octagonal pyramidal* numbers are the sums of consecutive octagonal numbers.

There are unlimited numbers of heptagonal and octagonal pyramidal numbers.

Star Numbers and Star Pyramidal Numbers

Octagonal numbers are also called *star* or *stellate* numbers because the dots corresponding to octagonal numbers can be arranged in the form of a four-pointed star.

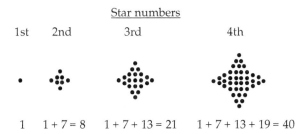

Star numbers

1st	2nd	3rd	4th
1	$1 + 7 = 8$	$1 + 7 + 13 = 21$	$1 + 7 + 13 + 19 = 40$

From the star pattern it can be shown that, allowing 0 to be a triangular number, each star number is the sum of a square number and four triangular numbers.

EXAMPLE

Find the value of the 2nd and 3rd star numbers.

2nd star number:
2nd square number $+ 4 \times$ 1st triangular number $= 4 + 4(1) = 8$

3rd star number:

3rd square number $+ 4 \times$ 2nd triangular number $= 9 + 4(3) = 21$

Star pyramidal numbers are the sums of consecutive star numbers.

The base and layers of a star pyramidal piling are star shaped. There is an unlimited number of star pyramidal numbers.

EXERCISE 71

1. Using the pattern begun in the example above, find the value of the star numbers from the 4th star number through the 8th star number.
2. Look at the patterns in the table below. Then extend the table through the 10th star pyramidal number.

Star pyramidal number	Value
1st	1
2nd	$1 + 8 = 9$
3rd	$9 + 21 = 30$

3. Make some star pyramidal pilings using translucent marbles and glue. Extend the table below through the 8th layer to help in planning your models.

Layer of model	Number of marbles in this layer	Total marbles in model thus far
1	1	1
2	$8 = 4 + 4(1)$	9
3	$21 = 9 + 4(3)$	30

4. Determine a formula that gives the star pyramidal numbers.

Rectangular Pyramidal Numbers

Recall that the first rectangular number was represented by an array of two dots. Other rectangular numbers were found by adding consecutive even numbers starting with 4.

Rectangular numbers

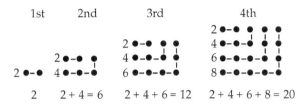

1st	2nd	3rd	4th
2	2 + 4 = 6	2 + 4 + 6 = 12	2 + 4 + 6 + 8 = 20

The concept of rectangular numbers was then extended to include numbers of the form $n(n + m)$ with $n = 1, 2, 3, \ldots$ and $m = 2, 3, 4, \ldots$.

EXAMPLE

Find the first four rectangular numbers of the form $n(n + m)$ with $m = 2$ and $n = 1, 2, 3, 4$.

The four numbers are

$$1(1 + 2) = 3 \qquad 2(2 + 2) = 8 \qquad 3(3 + 2) = 15 \qquad 4(4 + 2) = 24$$

Or, start with an array of 3 dots. Then, by adding consecutive odd numbers beginning with 5, other rectangular numbers in this pattern are formed.

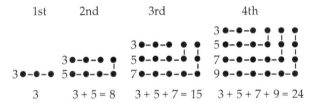

1st	2nd	3rd	4th
3	3 + 5 = 8	3 + 5 + 7 = 15	3 + 5 + 7 + 9 = 24

Several rectangular pyramidal piling layers are shown below.

2-type	3-type	4-type	5-type

Layer 1:

2	3	4	5

Layer 2:

6	8	10	12

The solid result of combined rectangular pilings whose top layer contains 4 spheres and whose bottom layer contains 10 spheres is shown on p. 124. Observe that the solid has five faces, a rectangular base face and four slanted faces.

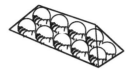

> *Rectangular pyramidal* numbers are the sums of consecutive rectangular numbers.

Different families of rectangular pyramidal numbers can be formed with different first layers of 2, 3, 4, and 5 spheres in the top layer as shown in the table that follows.

Rectangular pyramidal number type	Rectangular pyramidal number families
2-type (m = 1)	Sums of the consecutive rectangular numbers 2, 6, 12, 20, ...
3-type (m = 2)	Sums of the consecutive rectangular numbers 3, 8, 15, 24, ...
4-type (m = 3)	Sums of the consecutive rectangular numbers 4, 10, 18, 28, ...
5-type (m = 4)	Sums of the consecutive rectangular numbers 5, 12, 21, 32, ...

The number of rectangular pyramidal numbers of all types is unlimited.

EXERCISE 72

1. List the first 25 rectangular numbers starting with 3.
2. Starting with the rectangular number 4, other rectangular numbers can be formed as illustrated.

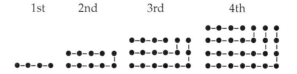

a) Label each array with an appropriate sum.
b) Starting with 4, describe how you can find the other rectangular numbers.
c) List the first 25 rectangular numbers starting with 4.

3. Starting with the rectangular number 5, other rectangular numbers can be formed as illustrated.

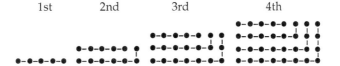

1st   2nd   3rd   4th

a) Label each array with an appropriate sum.

b) Starting with 5, describe how you can find the other rectangular numbers.

c) List the first 25 rectangular numbers starting with 5.

4. Make some rectangular pyramidal models of the 2-, 3-, 4-, and 5-type using marbles and glue. Extend each of the tables below through the 6th layer to help in planning your models.

	2-type			3-type	
Layer	Number of marbles in layer	Total marbles	Layer	Number of marbles in layer	Total marbles
1	2	2	1	3	3
2	6	8	2	8	11

	4-type			5-type	
Layer	Number of marbles in layer	Total marbles	Layer	Number of marbles in layer	Total marbles
1	4	4	1	5	5
2	10	14	2	12	17

5. Determine formulas that give the 2-, 3-, 4-, and 5-type rectangular pyramidal numbers.

Cubic Numbers

Cubic numbers are another example of solid or space numbers. The basic model here is the unit cube, a cube with edges of 1 unit in length. Unit cubes are then joined to form larger cubes.

The first cubic number is $1 \times 1 \times 1 = 1^3 = 1$. This is represented by 1 unit cube.

The second cubic number is $2 \times 2 \times 2 = 2^3 = 8$. This is represented by 8 unit cubes.

> *Cubic* numbers are the third powers of integers. The value of the nth cubic number is n^3.

The number of cubic numbers is unlimited.

EXERCISE 73

1. Find the value of the cubic numbers from the 3rd cubic number through the 10th cubic number.

Integers as Sums and Differences of Cubic Numbers, 1729

Integers can be written as sums of cubic numbers.
Sums of one cubic number and repetitions of cubic numbers are permitted.

EXAMPLE

Write each of the first ten integers as a sum of cubic numbers.

$$1 = 1^3$$
$$2 = 1^3 + 1^3$$
$$3 = 1^3 + 1^3 + 1^3$$
$$4 = 1^3 + 1^3 + 1^3 + 1^3$$
$$5 = 1^3 + 1^3 + 1^3 + 1^3 + 1^3$$
$$6 = 1^3 + 1^3 + 1^3 + 1^3 + 1^3 + 1^3$$
$$7 = 1^3 + 1^3 + 1^3 + 1^3 + 1^3 + 1^3 + 1^3$$
$$8 = 2^3$$
$$9 = 2^3 + 1^3$$
$$10 = 2^3 + 1^3 + 1^3$$

It has been shown that every integer can be written as the sum of at most nine cubic numbers.

Some cubic numbers can be written as a sum of cubic numbers in more than one way.

EXAMPLE

Write 27 as a sum of cubic numbers.

$$27 = 3^3$$
$$27 = 2^3 + 2^3 + 2^3 + 1^3 + 1^3 + 1^3$$

Srinivasa Ramanujan (1887–1920) grew up in Kumbakonam, Southern India. Ramanujan, in spite of a limited formal education and a short span of life, left behind a legacy of 4000 original theorems. His mathematical abilities were encouraged by the British mathematician, Godfrey Hardy (1877–1947). Once when Ramanujan was hospitalized in England, Hardy visited him and explained he had arrived in a cab with the dull number, 1729. Ramanujan immediately responded that 1729 was interesting as it is the first number that can be expressed as the sum of two cubes in two ways:

$$1729 = 1^3 + 12^3 = 9^3 + 10^3$$

EXAMPLE

Verify that 4104 can be written as the sum of two cubic numbers in two different ways.

$$4104 = 2^3 + 16^3 = 9^3 + 15^3$$

EXAMPLE

Pairs of numbers can be found such that the differences of their cubes are the same. Show that (9, 1) and (12, 10) are two such pairs.

$$9^3 - 1^3 = 729 - 1 = 728 \qquad\qquad 12^3 - 10^3 = 1728 - 1000 = 728$$

Can you find any other pairs of numbers such that the differences of their cubes are equal?

EXERCISE 74

1. Write each integer below as a sum of cubes.

   a) 14   b) 20   c) 29   d) 35
   e) 43   f) 67   g) 76   h) 89

2. Create a list of all two-digit numbers that can be expressed as the sum of two cubes.

3. Write each of the following cubic numbers as a sum of *at least two* cubic numbers. Use the *least* number of cubic numbers possible in your sum.

a) $64 = 4^3$ b) $125 = 5^3$ c) $216 = 6^3$ d) $343 = 7^3$
e) $512 = 8^3$ f) $729 = 9^3$ g) $1000 = 10^3$ h) $1331 = 11^3$
i) $1728 = 12^3$ j) $2197 = 13^3$

4. Write each integer below as a sum of two cubic numbers in two different ways.

a) 13,832   b) 20,683   c) 32,832   d) 39,312   e) 40,033
f) 46,683   g) 64,232   h) 65,728   i) 110,656

5. Some consecutive integers can each be written as sums of two cubic numbers. Try writing the consecutive numbers 854 and 855 as sums of two cubic numbers.

6. Some consecutive *even* integers can each be written as the sum of two cubic numbers. Try writing the consecutive even numbers 728 and 730 as sums of two cubic numbers. Find another pair of consecutive even integers with this property.

Pythagorean Parallelepiped Numbers

A *quadrilateral* is a simple closed figure whose sides are four straight-line segments. A *parallelogram* is a quadrilateral in which both pairs of opposite sides are parallel and congruent. A *rectangle* is a parallelogram with four right angles. The figure below shows a *rectangular solid* or *rectangular parallelepiped*. It has six rectangular faces. A *diagonal* of a rectangular solid is a line segment whose endpoints are two vertices that do not lie on the same face.

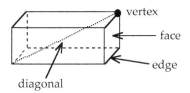

Pythagorean triples have been defined as triples of integers such that the sum of the squares of the two smaller numbers equals the square of the third number. For instance, (6, 8, 10) is a Pythagorean triple because $6^2 + 8^2 = 10^2$.

Recall that if the lengths of the sides of a triangle are a Pythagorean triple, then the triangle is a right triangle. The length of the hypotenuse, the side opposite the right angle, equals the square root of the sum of the squares of the lengths of the other two sides.

EXAMPLE

Given a right triangle with two sides of lengths 5 and 12. Find the length of the hypotenuse.

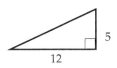

$$\sqrt{5^2 + 12^2} = \sqrt{169} = 13$$

So the hypotenuse is 13. Check that $(5, 12, 13)$ is a Pythagorean triple:

$$5^2 + 12^2 = 25 + 144 = 169 = 13^2$$

A *Pythagorean quadruple* is a quadruple of integers such that the sum of the squares of three of the numbers equals the square of the fourth number. Such quadruples are also called *Pythagorean parallelepiped numbers.*

EXAMPLE

Show that $(3, 6, 6, 9)$ and $(2, 3, 6, 7)$ are Pythagorean quadruples.

In each case, show that the sum of the squares of three of the numbers equals the square of the fourth number.

$$3^2 + 6^2 + 6^2 = 9 + 36 + 36 = 81 = 9^2$$
$$2^2 + 3^2 + 6^2 = 4 + 9 + 36 = 49 = 7^2$$

The rectangular parallelepiped that follows illustrates the geometric meaning of the Pythagorean quadruple $(2, 3, 6, 7)$. Here, the dimensions of the solid are 2 by 3 by 6. Notice that triangle ABC is a right triangle. The length of the diagonal AC is $\sqrt{6^2 + 2^2}$. Triangle ACD is also a right triangle. The length of the diagonal AD is $\sqrt{6^2 + 2^2 + 3^2} = \sqrt{49} = 7$. The largest number in a Pythagorean quadruple represents the length of the diagonal of a rectangular parallelepiped while the other three numbers represent the length of the edges of the solid.

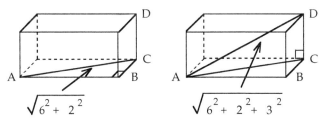

No more than two numbers are the same in a Pythagorean quadruple. In instances where the first three of the numbers in a quadruple *are* the same, the corresponding rectangular parallelepiped is a cube. However, if the side length of a cube is an integer, the length of its diagonal is not an integer. Thus, such a quadruple is *not* Pythagorean. For example, check that a cube with side length 2 has a diagonal with length $2\sqrt{3}$.

One way to find Pythagorean quadruples is to use Pythagorean triples. Find pairs of triples in which one number appears twice.

EXAMPLE

> Show that (3, 4, 12, 13) is a Pythagorean quadruple.
> Now, $3^2 + 4^2 = \underline{5^2}$ and $\underline{5^2} + 12^2 = 13^2$. Substituting $3^2 + 4^2$ for 5^2 in the second equation, we have $3^2 + 4^2 + 12^2 = 13^2$. So (3, 4, 12, 13) is a Pythagorean quadruple.

> The number of Pythagorean parallelepiped numbers is unlimited.

EXERCISE 75

> Complete each of the following quadruples so that the result is a Pythagorean quadruple.

> 1. (1, _, _, 9)   2. (2, _, _, 7)
> 3. (2, 4,_ ,_)   4. (2, 6, _, _)

> Try the **Investigation: Crisscross Cubes**.

Chapter 5

More Prime Connections

There is no inquiry which is not finally reducible to a question of Numbers ...
A. Comti (1798–1857)

Primes are the atoms of integers and over the centuries many fascinating facts and conjectures have been collected about them. A sample of these developments appears in this section.

Goldbach's Conjectures

In June 1742, Christian Goldbach's correspondence with Leonhard Euler led to his famous conjectures about prime numbers. They are

1) every even integer greater than 2 can be expressed as the sum of *two* primes (strong conjecture);
2) every odd integer greater than 5 can be expressed as the sum of *three* primes (weak conjecture as it is implied by (1)).

Although Goldbach took 1 to be a prime, 1 is not generally considered a prime, so the wording of Goldbach's conjectures are modified above and in the examples below. As of the present, significant progress has been made towards proving the conjectures, particularly, (2).

EXAMPLE

It appears that every even number greater than 2 is the sum of two primes, under the condition that primes may be repeated in the sum.

$$4 = 2 + 2 \qquad 6 = 3 + 3 \qquad 8 = 3 + 5$$
$$10 = 3 + 7 = 5 + 5 \qquad 12 = 5 + 7 \qquad 14 = 3 + 11 = 7 + 7$$

EXAMPLE

It appears that every even number greater than 4 is the sum of two *odd* primes.

$$16 = 13 + 3 = 11 + 5 \qquad\qquad 18 = 13 + 5 = 11 + 7$$

EXAMPLE

It appears that every even number greater than 6 is the sum of two *distinct odd primes*.

$$20 = 17 + 3 = 13 + 7 \qquad\qquad\qquad 22 = 19 + 3 = 17 + 5$$

Conjectures about primes and even numbers have been noted. What about primes and odd numbers?

EXAMPLE

It appears that every odd number greater than 5 is the sum of three primes. In fact, it appears that every odd number greater than 7 is the sum of three odd primes.

$$7 = 2 + 2 + 3 \qquad\qquad 9 = 2 + 2 + 5 = 3 + 3 + 3$$
$$11 = 2 + 2 + 7 = 3 + 3 + 5 \qquad 13 = 3 + 3 + 7$$

EXERCISE 76

1. Write each even number below as the sum of two primes. Primes may be repeated in the sum.

a) 24	b) 32	c) 46	d) 50	e) 68
f) 100	g) 110	h) 254	i) 298	j) 300
k) 456	l) 480	m) 512	n) 550	o) 600

2. Write each even number below as a sum of two distinct odd primes.

a) 26	b) 38	c) 44	d) 52	e) 60
f) 98	g) 102	h) 108	i) 234	j) 302
k) 422	l) 496	m) 506	n) 584	o) 606

3. Write each odd number below as a sum of three primes. Primes may be repeated in the sum.

a) 15	b) 27	c) 31	d) 53	e) 69
f) 101	g) 151	h) 253	i) 297	j) 301
k) 433	l) 487	m) 511	n) 595	o) 611

4. Check to see if it is possible to write each of the square numbers between 1 and 20^2 as the sum of two primes.

5. Using addition and subtraction, write primes in terms of primes. In each case supply $+$ or $-$ to create an equality.
   Solutions for 3 and 5 are given.

   $$3 = 1 + 2$$
   $$5 = 1 - 2 + 3 + 3$$
   $$7 = 1 \quad 2 \quad 3 \quad 5$$

```
11 = 1   2   3   5   7    7
13 = 1   2   3   5   7    11
17 = 1   2   3   5   7    11   13   13
19 = 1   2   3   5   7    11   13   17
23 = 1   2   3   5   7    11   13   17   19   19
29 = 1   2   3   5   7    11   13   17   19   23
```

Integers as Sums of Odd Integers

The Goldbach conjectures are the basis for the following explorations.

EXAMPLE

Write each even integer from 2 through 10 as the sum of two odd integers in as many ways as possible.

$2 = 1 + 1$ $4 = 1 + 3$ $6 = 1 + 5 = 3 + 3$
$8 = 1 + 7 = 3 + 5$ $10 = 1 + 9 = 3 + 7 = 5 + 5$

Thus, 2 and 4 can be written as the sum of two odds in one way, 6 and 8 can be written as the sum of two odds in two ways, and 10 can be written as the sum of two odds in three ways.

EXAMPLE

Write each odd integer from 3 through 9 as the sum of three odd integers.

$3 = 1 + 1 + 1$
$5 = 1 + 1 + 3$
$7 = 1 + 1 + 5 = 1 + 3 + 3$
$9 = 1 + 1 + 7 = 1 + 3 + 5 = 3 + 3 + 3$

EXAMPLE

Write each even integer from 2 through 10 as a sum of two distinct odd integers in as many ways as possible.

2, not possible $4 = 1 + 3$ $6 = 1 + 5$
$8 = 1 + 7 = 3 + 5$ $10 = 1 + 9 = 3 + 7$

EXAMPLE

Write each odd integer from 1 through 15 as a sum of three distinct odd integers in as many ways as possible.

1, 3, 5, 7 not possible.
 $9 = 1 + 3 + 5$ $11 = 1 + 3 + 7$
$13 = 1 + 3 + 9 = 1 + 5 + 7$ $15 = 1 + 3 + 11 = 3 + 5 + 7$

EXERCISE 77

1. Write each of the even integers from 12 through 46 as the sum of two odd integers in as many ways as possible.
2. Can you predict the number of ways other even integers can be written as the sum of two odd integers? If so, describe how.
3. Write each of the even integers from 12 through 30 as the sum of two distinct odd numbers in as many ways as possible.
4. Can you predict the number of ways other even integers can be written as the sum of two distinct odd integers? If so, describe how.
5. **Challenge** Determine the number of ways each of the odd integers from 11 through 39 can be written as a sum of three odd integers.
6. **Challenge** Can you predict the number of ways other odd integers can be written as the sum of three odd integers? If so, describe how.
7. **Challenge** Determine the number of ways each of the odd integers from 17 through 35 can be written as the sum of three *distinct* odd integers.
8. **Challenge** Can you predict the number of ways other odd integers can be written as the sum of three *distinct* odd integers? If so, describe how.

Integers as Sums of Two Composite Numbers

In Goldbach's conjectures, odd and even integers are expressed as sums of prime numbers. Can integers be written as sums of composite numbers? If so, what is the least number of composite numbers needed in the sums?

EXAMPLE

First check to see that it is not possible to write any integer from 1 through 7 as the sum of two composite numbers. Then note that

$12 = 4 + 8 = 6 + 6$ $13 = 4 + 9$ $14 = 4 + 10 = 6 + 8$
$15 = 6 + 9$ $16 = 4 + 12$
                             $\quad\ = 6 + 10 = 8 + 8$

EXERCISE 78

1. Write each of the integers from 17 through 26 as the sum of two composite numbers in as many ways as possible.

Positive Prime Pair Numbers

The following is a simple but challenging activity using prime numbers. It is related to a previous activity with positive square pair numbers (p. 89).

A number n is a *positive prime pair number* if it is possible to pair off the numbers from 1 to n with the numbers 1 to n so that the sum of each pair is a prime.

EXAMPLE

Let $A = \{1\}$ and $B = \{1\}$. Sets A and B contain the same element 1. Is it possible to pair the elements of A and B so that the sum is a prime number?

Pairing the number in set A with the number in set B and computing the sum you get $1 + 1 = 2$. Since 2 is a prime number, you can say that 1, the largest number in each set, is a *positive prime pair* number.

EXAMPLE

Is 2 a positive prime pair number?
Consider the sets $A = B = \{1, 2\}$.

elements from A:	1	2
elements from B:	$+2$	$+1$
	$\overline{3}$	$\overline{3}$

Since 3 is a prime, then for this choice of A and B, the pairing is possible. Therefore 2, the largest number in sets A and B is a positive prime pair number.

EXAMPLE

Is 3 a positive prime pair number?
Consider the sets $A = B = \{1, 2, 3\}$.

elements from A:	1	2	3
elements from B:	$+1$	$+3$	$+2$
	$\overline{2}$	$\overline{5}$	$\overline{5}$

Since 2 and 5 are both prime numbers, 3 is a positive prime pair number.

EXERCISE 79

1. Determine whether or not each of the numbers below is a positive prime pair number. Show the pairing of the numbers in sets A and B.
   a) 4; $A = B = \{1, 2, 3, 4\}$
   b) 5; $A = B = \{1, 2, 3, \dots , 5\}$
   c) 6; $A = B = \{1, 2, 3, \dots , 6\}$
   d) 7; $A = B = \{1, 2, 3, \dots , 7\}$
   e) 8; $A = B = \{1, 2, 3, \dots , 8\}$
   f) 9; $A = B = \{1, 2, 3, \dots , 9\}$
   g) 10; $A = B = \{1, 2, 3, \dots , 10\}$
   h) 11; $A = B = \{1, 2, 3, \dots , 11\}$
   i) 12; $A = B = \{1, 2, 3, \dots , 12\}$

2. How many more positive prime pair numbers can you find? Can you describe a procedure for making the pairings so that the sum of each pair in the group is a prime number?

Try the **Investigation: Prime Magic.**

Prime Line and Prime Circle Numbers

Two more activities that require pairings to produce sums that are prime are given in this section.

EXAMPLE

Take the set of consecutive integers $\{1, 2, 3, 4, 5\}$. Arrange them in a line such that the sum of every pair of adjacent integers is a prime.

One solution is: 5   2   3   4   1
That is, $5 + 2 = 7$, $2 + 3 = 5$, $3 + 4 = 7$, and $4 + 1 = 5$, all prime sums. Here 5 is a *prime line number* because 5 is the largest number in the given set of consecutive integers. Solutions may not be unique. For example, 1 4 3 2 5 and 5 2 1 4 3 are also acceptable arrangements.

EXAMPLE

Take the set of consecutive integers $\{1, 2, 3, 4\}$. Arrange them in a circle such that the sum of every pair of adjacent integers is a prime.

One solution is

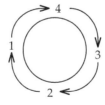

That is, $4 + 3 = 7$, $3 + 2 = 5$, $2 + 1 = 3$, and $1 + 4 = 5$ are all prime sums. The number 4 is a *prime circle number* because 4 is the largest number in the given set of consecutive integers.

Sometimes it is more convenient to arrange the integers in a row such that
a) the sum of every pair of adjacent integers is a prime, and
b) the sum of the first and last integer in the row is also a prime.

Thus, for the prime circle number 4, here are some of the possible linear arrangements.

3 2 1 4   2 1 4 3   1 4 3 2

EXERCISE 80

1. Which of the following numbers is a prime line number? Show the linear arrangement of the set of consecutive integers.

   a) 3; {1, 2, 3}   b) 7; {1, 2, 3, ... , 7}
   c) 9; {1, 2, 3, ... , 9}   d) 12; {1, 2, 3, ... , 12}
   e) 13; {1, 2, 3, ... , 13}   f) 14; {1, 2, 3, ... , 14}
   g) 15; {1, 2, 3, ... , 15}   h) 16; {1, 2, 3, ... , 16}
   i) 17; {1, 2, 3, ... , 17}   j) 18; {1, 2, 3, ... , 18}
   k) 19; {1, 2, 3, ... , 19}   l) 20; {1, 2, 3, ... , 20}

2. Show that each of the following numbers is a prime circle number by giving a linear arrangement of the set of consecutive integers.

   a) 2; {1, 2}   b) 6; {1, 2, 3, ... , 6}
   c) 8; {1, 2, 3, ... , 8}   d) 10; {1, 2, 3, ... , 10}
   e) 12; {1, 2, 3, ... , 12}   f) 14; {1, 2, 3, ... , 14}
   g) 16; {1, 2, 3, ... , 16}

3. Formulate some conjectures about prime line and prime circle numbers.
4. Complete the Prime Line Triangle. The first and last numbers in all rows are given. In row n each of the numbers 1 through n must appear. Also, the numbers must be placed in each row so that the sum of adjacent pairs is a prime.

   a. Study rows 2 through 6, which are completed and then fill in rows 7 through 15.
   b. Try extending the triangle beyond row 15.
   c. Describe any patterns you observe.

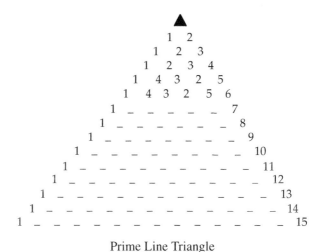

Prime Line Triangle

Beprisque Numbers

> A *beprisque* number is a number in a sequence of consecutive integers that is adjacent to a prime number and a square number, or a number adjacent to a square number and a prime number.

The name "beprisque" comes from the initial letters of "between prime and square" with an extra "e" added. Possibly, the number of beprisque numbers is unlimited.

EXAMPLE

Show that 3, 8, and 10 are beprisque numbers.

In the triple 2, 3, 4, the number 3 is adjacent to the prime 2 and the square 4.

In the triple 7, 8, 9, the number 8 is adjacent to the prime 7 and the square 9.

In the triple 9, 10, 11, the number 10 is adjacent to the square 9 and the prime 11.

Using the definition above and the list of square numbers, develop an efficient strategy for locating beprisque numbers. Employ your strategy and verify the first five beprisque numbers are 2, 3, 8, 10, and 24.

EXERCISE 81

1. List all beprisque numbers less than 200.
2. 3 is an odd beprisque number. Are there any other odd beprisque numbers? Give a reason for your answer.
3. List the integers from 2 through 16. Indicate those numbers that are beprisque numbers or that can be written as the sum of beprisque numbers. Use the least number of beprisque numbers in each sum.
4. Continue problem 3 with larger numbers.

A Primes-Between Property

Paul Erdös (1913–1996) of Budapest, Hungary published his first results in number theory at the age of 18.

Mathematics clearly was his life. In fact, he became a mathematical nomad who wandered about the world visiting, teaching and collaborating. By the time he died he had already authored or co-authored over 1500 papers. He has been identified as one of the 20th century's greatest mathematicians.

At age 20, Erdös discovered an elegant proof of a property about primes that had first been proved by the Russian mathematician, P. L. Chebyshev (1821–1894), in the 19th century. The property stated that:

For every $n > 1$, there is at least one prime between n and $2n$.

Germain Primes

Sophie Germain (1776–1831) of France overcame many obstacles to become an outstanding mathematician. For a time, she wrote under the pseudonym, Monsieur LeBlanc, as she felt that because she was a woman, her work would not be taken seriously. Yet, in time, the great German mathematician, Carl F. Gauss said of Sophie Germain " ... without a doubt she must have the noblest courage, quite extraordinary talents, and a superior genius." Germain investigated a property that applies to some prime numbers. The numbers now carry her name.

> A prime p for which $2p + 1$ is prime is a *Germain prime*.

EXAMPLE

Verify that the first three primes 2, 3, and 5 are Germain primes.

$$p \rightarrow 2(p) + 1$$
$$2 \rightarrow 2(2) + 1 = 5 \quad \text{a prime}$$
$$3 \rightarrow 2(3) + 1 = 7 \quad \text{a prime}$$
$$5 \rightarrow 2(5) + 1 = 11 \quad \text{a prime}$$

EXAMPLE

Verify that 7 is not a Germain prime.

7 is not a Germain prime because $2(7) + 1 = 15$, a composite number.

EXERCISE 82

1. Determine which two-digit primes are also Germain primes.
2. If possible, find a three-digit Germain prime that ends in

   a) 1   b) 3   c) 7   d) 9.

3. Divide each two-digit Germain prime by 4 and record its remainder. Do you notice any patterns? If so, describe them.
4. For each case in problem 3 where the remainder is 3, check to see if $2p + 1$ divides $2^p - 1$.
5. Build a chain of Germain primes starting with 2. Stop when the result is not a Germain prime. For example,

$$2 \rightarrow 2(2) + 1 = 5$$
$$5 \rightarrow 2(5) + 1 = 11$$
$$11 \rightarrow 2(11) + 1 = 23$$
$$23 \rightarrow 2(23) + 1 = 47$$
$$47 \rightarrow 2(47) + 1 = 95, \text{ not a prime}$$

So 47 is not a Germain prime.

The chain of Germain primes here is: $2 \rightarrow 5 \rightarrow 11 \rightarrow 23$.

Starting with a two-digit prime, display the longest chain of Germain primes you can find.

Sophie Germain went on to prove a special case of Fermat's famous Last Theorem. She proved that if p is a Germain prime greater than 2, then there are *no* integers x, y, z different from 0 and not multiples of p such that

$$x^p + y^p = z^p.$$

Twin Primes

> Two primes whose difference is two are *twin primes*.

EXAMPLE

Find all twin primes between 100 and 150.

Verify that they are:

101 and 103   107 and 109   137 and 139   149 and 151.

It is not known if the number of twin primes is infinite.

EXERCISE 83

1. List the twin primes less than 100.
2. Examine the sums of twin primes less than 100. Form a conjecture about the sum of twin primes.
3. Prove the conjecture you made in problem 2.
4. Examine the products of twin primes less than 100. Form a conjecture about the product of twin primes.
5. Prove the conjecture you made in problem 4.
6. Express each of the following numbers as the difference of two primes in 10 ways.

    a) 2   b) 4   c) 6

Semiprimes and Boolean Integers

> A *semiprime* number is a product of two distinct primes.

EXAMPLE

Find the first six semiprime numbers.

The numbers 6, 10, 14, 15, 21, and 22 are the first six semiprimes because

$6 = 2 \times 3$	$10 = 2 \times 5$	$14 = 2 \times 7$
$15 = 3 \times 5$	$21 = 3 \times 7$	$22 = 2 \times 11$

EXAMPLE

Express the integers from 24 through 27 as a sum of two or more semiprimes.

$24 = 10 + 14$	$25 = 10 + 15$
$26 = 6 + 6 + 14$	$27 = 6 + 21$

> A *boolean* integer is any integer whose prime factorization contains only distinct primes raised to the first power.

Boolean integers are named in honor of George Boole (1815–1864). Boole was a professor of mathematics at Queen's College in Cork, Ireland. He made significant contributions to mathematics by developing a structure for logic that came to be known as boolean algebra. Boolean integers are connected to boolean algebra. All primes and semiprimes are boolean integers.

EXAMPLE

List the first five boolean integers.

2, 3, 5, 6, and 7 are the first five boolean integers. Of these, 2, 3, 5, and 7 are primes while 6 is a semiprime.

EXAMPLE

What number is the smallest boolean integer that is not prime or semiprime?

This number must contain more than two different prime factors. Hence $2 \times 3 \times 5 = 30$ is the smallest such boolean integer.

EXERCISE 84

1. Write each of the integers from 28 through 48 as the sum of the least number of semiprimes. Semiprimes may be repeated in the sum.
2. Name five numbers that can be expressed as a sum of semiprimes in more than one way.
3. Name all boolean integers less than 100 that are not primes or semiprimes.
4. For each of the boolean integers below, list the divisors of the integer, indicate how many distinct prime divisors each integer has, and give the number of divisors each integer has.

a) 3	b) 5	c) 6	d) 14	e) 15	f) 23
g) 30	h) 31	i) 33	j) 42	k) 59	l) 66

5. Using the results of problem 4, make a conjecture about the number of divisors of a boolean integer.

Snowball Primes

The collection of numbers known as "snowball primes" have been studied under different names and descriptions in number theory. Professor Leslie Card (1893–1986) of the University of Illinois used this label and definition that follows in 1968. He was a professor of animal science and recreational mathematics was his hobby.

> *Snowball primes* are a collection of prime numbers such that each number in the list after the first is formed by adding a single-rightmost digit to the preceding number.

While investigating snowball primes, *assume that 1 is a prime.*

EXAMPLE

Start with 1.

$1 \rightarrow 19 \rightarrow 193 \rightarrow 1933 \rightarrow 19333 \rightarrow 193337$ are each primes.

This is a six-element set of snowball primes. It is not possible to make this set larger because the numbers 1933371, 1933373, 1933377, and 1933379 are all composite.

EXERCISE 85

Starting with each of the following sets of snowball primes, how *large* a set of snowball primes can you make?

1. $1 \rightarrow 13 \rightarrow$ 2. $1 \rightarrow 19 \rightarrow$ 3. $2 \rightarrow 23 \rightarrow$ 4. $2 \rightarrow 29 \rightarrow$
5. $3 \rightarrow 37 \rightarrow$ 6. $5 \rightarrow 59 \rightarrow$ 7. $7 \rightarrow 71 \rightarrow$

Lucky Numbers

Lucky numbers were the brainchild of Stanislaw Ulam (1909–1984), an accomplished Polish–American mathematician who first came to the USA in 1935, and a group of scientists at Los Alamos National Laboratory, New Mexico around 1956.

These numbers have many properties that are very similar to those of the prime numbers.

Lucky numbers are found by a sieve process similar to the sieve method the Greek mathematician, Eratosthenes, used to find prime numbers.

Algorithm to find lucky numbers

1. Write the integers from 1 to as high as you wish. For the purpose of illustration 1 through 50 are used.
2. Call 1 a lucky number. Circle it. The next number not circled is 2. Starting with 1, cross out every second number.

①2̸ 3 4̸ 5 6̸ 7 8̸ 9 1̸0̸ 11 1̸2̸ 13

1̸4̸ 15 1̸6̸ 17 1̸8̸ 19 2̸0̸ 21 2̸2̸ 23 2̸4̸

25 2̸6̸ 27 2̸8̸ 29 3̸0̸ 31 3̸2̸ 33 3̸4̸ 35

3̸6̸ 37 3̸8̸ 39 4̸0̸ 41 4̸2̸ 43 4̸4̸ 45 4̸6̸

47 4̸8̸ 49 5̸0̸

This action eliminates all the even numbers. The numbers that remain are

① 3 5 7 9 11 13 15 17 19 21 23

25 27 29 31 33 35 37 39 41 43 45

47 49

3. Now circle the next number left after 1, that is, circle 3. Call 3 a lucky number. Starting with 1, cross out every third number in the list.

①③ 5̸ 7 9 1̸1̸ 13 15 1̸7̸ 19 21 2̸3̸

25 27 2̸9̸ 31 33 3̸5̸ 37 39 4̸1̸ 43 45

4̸7̸ 49

The numbers that remain now are

①③ 7 9 13 15 19 21 25 27 31 33

37 39 43 45 49

4. Now circle the next number left after 3, that is, circle 7. So 7 is a lucky number. Starting with 1, cross out every seventh number in the list.

①③⑦ 9 13 15 1̸9̸ 21 25 27 31 33

37 3̸9̸ 43 45 49

The numbers that remain now are

①③⑦ 9 13 15 21 25 27 31 33 37

43 45 49

5. Now circle the next number left after 7, that is, circle 9. So 9 is lucky. Starting with 1, cross out every ninth number in the list. The number 27 is then removed.

6. Now circle the next number left after 9, that is, circle 13. So, 13 is lucky. Starting with 1, cross out every thirteenth number in the list. The number 45 is then removed.

7. Now circle the next number left after 13, that is, circle 15. So 15 is lucky. Starting with 1, cross out every fifteenth number in the list. Since there are only 13 numbers left, no number is removed. All the numbers that remain are lucky. The lucky numbers less than 50 are:

$$1, 3, 7, 9, 13, 15, 21, 25, 31, 33, 37, 43, \text{ and } 49.$$

This procedure can be used to find the lucky numbers less than 100, 200, and so on. There are an infinite number of lucky numbers but it is not known if there are an infinite number of *prime* luckies. The first 60 lucky numbers are given in the table below.

		Lucky numbers			
1	37*	87	141	205	267
3*	43*	93	151*	211*	273
7*	49	99	159	219	283*
9	51	105	163*	223*	285
13*	63	111	169	231	289
15	67*	115	171	235	297
21	69	127*	189	237	303
25	73*	129	193*	241*	307*
31*	75	133	195	259	319
33	79*	135	201	261	321

*The starred numbers are primes.

Prime and Lucky Numbers

A few instances are singled out that illustrate the similarity of properties between prime and lucky numbers. In the examples in this section, three conjectures about lucky numbers are explored that are similar to the Goldbach conjectures regarding primes.

EXAMPLE

It appears that every even integer is the sum of two lucky numbers.

$$2 = 1 + 1 \qquad 4 = 1 + 3 \qquad 6 = 3 + 3 \qquad 8 = 1 + 7$$

EXAMPLE

It appears that every even integer greater than 6 is the sum of two *distinct* lucky numbers.

$$8 = 1 + 7 \qquad 10 = 1 + 9 \qquad 12 = 3 + 9 \qquad 14 = 1 + 13$$

EXAMPLE

It appears that every odd integer greater than 1 is the sum of three lucky numbers.

$$3 = 1 + 1 + 1 \qquad\qquad\qquad 5 = 1 + 1 + 3$$
$$7 = 1 + 3 + 3 \qquad\qquad\qquad 9 = 1 + 1 + 7$$

EXERCISE 86

1. Write each even number below as a sum of two lucky numbers.

a) 16	b) 18	c) 20	d) 22	e) 24
f) 32	g) 48	h) 56	i) 64	j) 70
k) 82	l) 96	m) 104	n) 212	o) 318

2. Write each even number below as a sum of two *distinct* lucky numbers.

a) 26	b) 28	c) 30	d) 34	e) 36
f) 38	g) 46	h) 52	i) 60	j) 78
k) 88	l) 94	m) 102	n) 216	o) 300

3. Write each odd number below as a sum of three lucky numbers.

a) 13	b) 15	c) 17	d) 19	e) 21
f) 33	g) 45	h) 57	i) 61	j) 79
k) 81	l) 99	m) 103	n) 215	o) 307

4. Investigate other ways to compare primes and lucky numbers.

Polya's Conjecture about Odd- and Even-Type Integers

In 1940 George Polya (1887–1985) came to the USA from Budapest, Hungary after studying in Germany and holding a teaching position in Zurich, Switzerland. Polya was a highly respected and talented mathematician, but his major legacy to the world community of mathematics teachers and students was his work and writings about mathematical problem solving. His insights and suggestions were conveyed with clarity and enthusiasm. Included in Polya's research are conjectures involving odd- and even-type integers.

> An *odd-type* integer is a positive integer whose prime factorization contains an odd number of primes, including repetitions.

All primes are odd-type integers because their factorizations contain only one prime.

EXAMPLE

Show that 8 and 27 are odd-type integers.

$8 = 2 \times 2 \times 2$ $27 = 3 \times 3 \times 3$

The prime factorizations of 8 and 27 each contain three primes. Hence, they are odd-type integers.

> An *even-type* integer is a positive integer whose prime factorization contains an even number of primes, including repetitions.

The integer 1 is neither odd-type nor even-type.

EXAMPLE

Show that 4, 6, and 24 are even-type integers.

$4 = 2 \times 2$ $6 = 2 \times 3$ $24 = 2 \times 2 \times 2 \times 3$

The prime factorizations of 4 and 6 contain two primes and the factorization of 24 contains four prime factors, so 4, 6, and 24 are even-type integers.

In 1919, Polya conjectured that the number of even-type integers never equals the number of odd-type integers in any interval beginning with 2. Other mathematicians have discovered some interesting facts while investigating this conjecture.

EXERCISE 87

1. Find the number of odd-type and even-type integers in each of the intervals given below.

   a) 2 through 10   b) 11 through 20
   c) 21 through 30   d) 31 through 40
   e) 41 through 50   f) 51 through 60
   g) 61 through 70   h) 71 through 80
   i) 81 through 90   j) 91 through 100

2. Use the results of problem 1 to find the total number of odd-type and even-type integers in each of the following intervals.

   a) 2 through 10   b) 2 through 20
   c) 2 through 30   d) 2 through 40
   e) 2 through 50   f) 2 through 60
   g) 2 through 70   h) 2 through 80
   i) 2 through 90   j) 2 through 100

Balanced Numbers

A number n is a *balanced* number if and only if there are as many composite numbers between 1 and n as there are prime numbers.

EXAMPLE

Is 8 a balanced number?

The numbers, 2, 3, 4, 5, 6, and 7 are between 1 and 8. There are 2 composite numbers, 4 and 6, but 4 prime numbers, 2, 3, 5, and 7, between 1 and 8. Therefore 8 is not balanced.

EXAMPLE

Consider 12. Is it balanced?

The numbers 2, 3, 4, 5, 6, 7, 8, 9, 10, and 11 are between 1 and 12. There are 5 composite numbers, 4, 6, 8, 9, and 10, as well as 5 prime numbers, 2, 3, 5, 7, and 11 between 1 and 12. Therefore 12 is balanced.

EXERCISE 88

1. Are there any single-digit balanced numbers? If so, which numbers are they?
2. Which of the following numbers, if any, are balanced?

    a) 14   b) 16   c) 18

3. Do you think the number of balanced numbers is unlimited? Explain your answer.

Fermat Numbers

Numbers of the form $2^{2^n} + 1$, $n \geq 0$, are called *Fermat* numbers. Fermat numbers are labeled F_0, F_1, F_2, ..., F_n.

EXAMPLE

Find the value of the Fermat numbers F_0 and F_1.

$$F_0 = 2^{2^0} + 1 = 2^1 + 1 = 3$$
$$F_1 = 2^{2^1} + 1 = 2^2 + 1 = 5$$

Note that F_0 and F_1 are prime numbers. Fermat made a conjecture that all numbers F_n are prime numbers.

There is an interesting application of Fermat primes. The Greeks knew constructions for regular polygons with N sides such that $N = 2^p(3)^j(5)^k$, with p a non-negative integer, and j, k a 0 or 1. In 1796, Carl Friedrich Gauss, at the age of 19, demonstrated that a regular polygon with 17 sides could be constructed. Then, in his famous work on number theory, **Disquisitiones Arithmeticae**, published in 1801, he proved the theorem: If the number of sides of a regular polygon is of the form

1. 2^p, for $p \geq 2$, or
2. 2^p times a product of *different* Fermat primes, $p \geq 0$,

then the regular polygon is constructible with straightedge and compass. He also stated but did not provide a proof of the converse of the theorem.

EXERCISE 89

1. Is each of the Fermat numbers from F_0 through F_4 prime? Complete the entries in the table. Remember, $2^{2^n} = 2^{(2^n)}$

Fermat numbers	Prime? yes/no	Prime factorization
$F_0 = 2^{2^0} + 1 = 3$	yes	
$F_1 = 2^{2^1} + 1 = 5$	yes	
$F_2 =$		
$F_3 =$		
$F_4 =$		
$F_5 = 2^{2^5} + 1 = 2^{32} + 1 = 4{,}294{,}967{,}297$	no	$641 \times 6{,}700{,}417$

2. The first two Fermat primes are 3 and 5. Regular polygons of 3, 4, and 5 sides can be constructed using a straightedge and compass. Use Gauss' conditions (1) and (2) to name the *next ten* regular polygons that are constructible with straightedge and compass.

Cullen Numbers

English Jesuit, James Cullen (1867–1933) first worked with this special set of numbers around 1905. In time, these numbers came to bear his name.

> Numbers of the form $n(2^n)+1$ are called *Cullen* numbers.

The first Cullen number is a prime since $1(2^1)+1 = 3$. Strangely, the *second prime* Cullen number does not occur until $n = 141$ and $141(2^{141}) + 1$ has 45 digits. The 16th *prime* Cullen number with $n = 6{,}679{,}881$ has 2,010,852 digits!

EXAMPLE

Show that 9, 25, and 65 are Cullen numbers.

Since 9, 25, and 65 can be written in the form $n(2^n) + 1$, they are Cullen numbers.

$$9 = 2(2^2) + 1 \qquad 25 = 3(2^3) + 1 \qquad 65 = 4(2^4) + 1$$

There are an unlimited number of Cullen numbers and it is conjectured that there are an infinite number of Cullen primes.

EXERCISE 90

1. Find the value of the Cullen numbers $n(2^n) + 1$ where $n = 2, \ldots, 15$.
2. Numbers of the form $n(2^n) - 1$ are called Cullen numbers of the second kind. Find these Cullen numbers for $n = 5, 10, 15, 20, 25$.
3. Study the numbers in problems 1 and 2. Describe any patterns you observe.

Ruth–Aaron Numbers

Ruth–Aaron numbers are named after the major league ballplayers George Herman (Babe) Ruth, Jr., and Henry (Hank) Aaron. On April 8, 1974 in Atlanta, Georgia, Aaron hit his 715th major league homerun and broke the record of 714 regular season homeruns held by Babe Ruth. Graduate student Carol Nelson, and professors David E. Penney and Carl Pomerance at the University of Georgia gave the name to special pairs of consecutive numbers. This was prompted by their work with prime factorization and the properties of consecutive numbers 714 and 715.

Recall that the prime factorization theorem states that each composite number N can be written

$$N = p_1^{a_1} p_2^{a_2} \cdots p_k^{a_k}$$

where the p_i's are prime numbers and the a_i's are the matching exponents for each prime.

Let $S(N) = a_1 p_1 + a_2 p_2 + \cdots + a_k p_k$.

Thus if $N = 60$,

then $N = 2^2 \cdot 3^1 \cdot 5^1$

and

$S(60) = (2 \cdot 2) + (1 \cdot 3) + (1 \cdot 5) = 12$.

The group discovered that $S(714) = S(715)$.

That is, $714 = 2^1 \cdot 3^1 \cdot 7^1 \cdot 17^1$

and $S(714) = (1 \cdot 2) + (1 \cdot 3) + (1 \cdot 7) + (1 \cdot 17) = 29$

while $715 = 5 \cdot 11 \cdot 13$

with $S(715) = (1 \cdot 5) + (1 \cdot 11) + (1 \cdot 13) = 29$

In general,

Two consecutive numbers n, $n + 1$ are a Ruth–Aaron pair of numbers if $S(n) = S(n+ 1)$, where $S(n)$ is the sum of the product of each prime factor of n and its corresponding exponent.

EXERCISE 91

1. Find all Ruth–Aaron pairs of consecutive numbers where each component in a pair is less than 100.
2. Find all pairs of consecutive numbers with each component less than 100 whose prime factor and exponent product sums differ by

   a) 1   b) 2   c) 3   d) 4

3. Name the largest $S(n)$ for $n < 100$.
4. For $n < 100$ give the most frequently occurring $S(n)$.

Chapter 6

Digital Patterns and Noteworthy Numbers

God made the integers, man made the rest.

Leopold Kronecker (1823–1891)

There are many number types that emerge that are based on relations involving the individual digits that compose the representation of the number. Some of these along with a variety of notable numbers are introduced in this section.

Monodigit and Repunit Numbers and Langford Sequences

> A *monodigit* number is a number that is formed using only one distinct digit.

The numbers 11, 222, and 33333 are examples of monodigit numbers.

> Monodigit numbers composed of 1s are called *repunit* numbers. The word repunit comes from *rep*eating *unit*.

The word *repunit* was created by Albert H. Beiler in 1966. It appeared in his book **Recreations in the Theory of Numbers: The Queen of Mathematics Entertains.** The first five repunit numbers are 1, 11, 111, 1111, and 11111.

Repunit numbers are generated by the formula $\dfrac{10^n - 1}{9}$.

If n is a composite number, the repunit number is also composite. If n is a prime number, the repunit number may or may not be a prime number.

EXAMPLE

Determine the status of the repunit numbers generated by $n = 2, 3, 4$.

If $n = 2$, then $\dfrac{10^2 - 1}{9} = \dfrac{(10 - 1)(10 + 1)}{9} = 1(11)$

Thus for the prime 2, the repunit number is a prime, 11.

If $n = 3$, then

$$\frac{10^3 - 1}{9} = \frac{(10 - 1)(10^2 + 10 + 1)}{9} = 1(111).$$

Thus, for the prime 3, the repunit number 111 is composite because $111 = 3(37)$.

If $n = 4$, then
$$\frac{10^4 - 1}{9} = \frac{(10^2 - 1)(10^2 + 1)}{9} = \frac{99(101)}{9} = 11(101).$$
Thus, for the composite number 4, the repunit number 1111 is also composite.

EXERCISE 92

1. Compute the square of each of the repunit numbers from 1 through 1,111,111,111.

2. Compute the following products.

   a) 10101×11 b) 1010101×11
   c) 101010101×11 d) 10101010101×11

3. Show that each expression below equals a composite number. Follow the pattern of the examples above.

   a) $\dfrac{10^6 - 1}{9}$ b) $\dfrac{10^8 - 1}{9}$ c) $\dfrac{10^9 - 1}{9}$ d) $\dfrac{10^{10} - 1}{9}$

4. Find the missing prime factorizations in the list of repunit numbers below.

$$11 = 11$$
$$111 = 3(37)$$
$$1111 = 11(101)$$
$$11,111 = 41(271)$$
$$111,111 = ?$$
$$1,111,111 = 239(4649)$$
$$11,111,111 = ?$$
$$111,111,111 = 32(37)(333667)$$

5. Notice that 7 is a prime and 7 divides six 1s. That is, $111,111 \div 7 = 15,873$. Also notice that 11 is a prime and 11 divides ten 1s. That is, $1,111,111,111 \div 11 = 101,010,101$.

   a) How many 1s does the prime 13 divide?
   b) How many 1s does the prime 17 divide?

Often mathematical problems are generated from very simple circumstances. C. Dudley Langford of England described in 1958 how Langford sequences came to be. He observed his young son playing with blocks and noticed in particular, certain patterns in the color arrangements in the stacks of blocks. These caused him to pose the generalization that is called a *Langford sequence*.

A (k, n) *Langford sequence* is a rearrangement of k copies of the integers from 1 to n such that between each occurrence of the integer j there are exactly j integers with $j = 1, 2, \ldots, n$.

EXAMPLE

If $k = 2$ and $n = 3$, find the Langford sequence.

It is all a matter of arrangement. Thus, the 6 given numbers 1, 1, 2, 2, 3, and 3, must be arranged so that:

> 1 number lies between two 1s
> 2 numbers lie between two 2s
> 3 numbers lie between two 3s

One solution would be 2 3 1 2 1 3.
Can you find another?

Langford's observation about such patterns has stimulated many investigations and extensions.

EXERCISE 93

1. If $k = 2$ and $n = 4$, find a Langford sequence.
2. If $k = 2$ and $n = 7$, find a Langford sequence.
3. If $k = 2$ and $n = 8$, find a Langford sequence.

Social and Lonely Numbers

Social numbers are not the same as the sociable numbers described in a previous section. Study the following examples carefully.

EXAMPLE

Consider the number 13 and the computation.

$$13 + (1 + 3) = 17$$

13 is called a *generator* of 17. Because 17 has a generator, 17 is a *social* number.

EXAMPLE

Consider the numbers 91 and 100.

$$91 + (9 + 1) = 101$$
$$100 + (1 + 0 + 0) = 101$$

91 and 100 are both generators of 101. Because 101 has at least one generator, 101 is a *social* number. Social numbers have at least one generator.

A number x that can be written as the sum of another number y and the digits of y is called a *social* number. The number y is called a *generator* of x.

Numbers without generators are called *lonely* numbers.

The numbers 1, 3, 5, 7, and 9 are examples of lonely numbers.

EXAMPLE

Is 15 a social number? A lonely number?

To find out, one strategy to use is to make a table.

Start with the sum of 15 and its digits and work backward.

$$15 + (1 + 5) = 21$$
$$14 + (1 + 4) = 19$$
$$13 + (1 + 3) = 17$$
$$12 + (1 + 2) = 15 \leftarrow$$

12 is a generator of 15, so 15 is a social number.

EXAMPLE

Is 20 a social number? A lonely number?

Make a table starting with the sum of 20 and its digits and work backward.

$$20 + (2 + 0) = 22$$
$$19 + (1 + 9) = 29$$
$$18 + (1 + 8) = 27$$
$$17 + (1 + 7) = 25$$
$$16 + (1 + 6) = 23$$
$$15 + (1 + 5) = 21$$

$$\leftarrow$$

$$14 + (1 + 4) = 19$$

The gap indicates 20 does not have a generator.

So 20 is a lonely number.

Each integer generates a social number. Since the number of integers is unlimited, it can be shown that the number of social numbers is unlimited. Do you think the number of lonely numbers is unlimited? Explain your answer.

EXERCISE 94

1. Generate a social number from each of the following starting numbers.

a) 1	b) 2	c) 3	d) 4
e) 5	f) 6	g) 7	h) 8
i) 9	j) 10	k) 11	l) 12
m) 92 and 101	n) 93 and 102	o) 94 and 103	

2. Which of the following numbers are social?

a) 32	b) 41	c) 42	d) 48
e) 53	f) 64	g) 69	h) 75
i) 77	j) 84	k) 86	l) 90
m) 97	n) 98	o) 108	p) 120

Additive Multidigital Numbers

The sum of the digits of a number leads to interesting patterns and divisibility rules. For instance, it is well known and can easily be shown that if the sum of the digits of a number is divisible by 3, then the number is divisible by 3.

EXAMPLE

Test to see if 12 is divisible by 3.

The sum of the digits in 12 is $1 + 2 = 3$. Since the sum of the digits is divisible by 3, 12 is also divisible by 3.

Another well-known divisibility rule states: if the sum of the digits of a number is divisible by 9, then the number is divisible by 9.

EXAMPLE

Test to see if 6309 is divisible by 9.

The sum of the digits in 6309 is $6 + 3 + 0 + 9 = 18$. Since the sum of the digits is divisible by 9 then 6309 is also divisible by 9.

EXERCISE 95

1. Determine whether or not each of the following numbers is divisible by 3 by testing the sum of its digits.

    a) 14   b) 27   c) 111
    d) 137   e) 4122   f) 48,323

2. Determine whether or not each of the following numbers is divisible by 9 by testing the sum of its digits.

    a) 41   b) 72   c) 459   d) 777
    e) 8926   f) 66,666   g) 12,345,678

3. Write a paragraph or two that explains why the divisibility patterns for 3 and 9 work.

Dattaraya Ramchandra Kaprekar (1905–1986) worked as a teacher in Devali, India schools for more than 30 years and did original work in recreational number theory for over 50 years. Kaprekar used the concept of summing the digits of a number and made this definition.

> An *additive multidigital* number is a number that is divisible by the sum of its digits.

The first three additive multidigital numbers with divisor 3 are 3, 12, and 21. The first three additive multidigital numbers with divisor 9 are 9, 18, and 27. The first three additive multidigital numbers with divisor 5 are 5, 50, and 140.

Since 10, and every integral power of 10, is an additive multidigital number with divisor 1, then it follows that the number of additive multidigital numbers with divisor 1 is unlimited. In fact, the number of additive multidigital numbers with any named divisor is unlimited.

EXERCISE 96

1. Which of the following numbers is an additive multidigital number? Give a reason for your choice.

   a) 8   b) 11   c) 42   d) 55   e) 63
   f) 75   g) 84   h) 114   i) 195   j) 198

2. For each divisor from 2 through 19, name a three-digit additive multidigital number with that divisor.

3. For each of the following divisors, find three additive multidigital numbers with that divisor.

   a) 3   b) 5   c) 7   d) 9   e) 11   f) 13
   g) 14   h) 15   i) 16   j) 17   k) 18   l) 19
   m) 21   n) 22   o) 25   p) 31   q) 37   r) 40

4. Study the following pattern. If possible, extend the pattern for the factorial numbers from 5! through 12! Factorial number n, symbolized as $n!$, equals the product of the consecutive integers from 1 through n.

   $1! = 1$ is an additive multidigital number with divisor 1.
   $2! = 2$ is an additive multidigital number with divisor 2.
   $3! = 6$ is an additive multidigital number with divisor 6.
   $4! = 24$ is an additive multidigital number with divisor 6.

Multiplicative Multidigital Numbers

Kaprekar's idea can be extended to the product of the digits of a number.

A *multiplicative multidigital* number is a number that is divisible by the product of its digits.

EXAMPLE

Is 36 a multiplicative multidigital number?

The digits of 36 have a product of $3 \times 6 = 18$.
Since 36 is divisible by 18, it follows that 36 is a multiplicative multidigital number.

The integers 1 through 9 are all multiplicative multidigital numbers. No number with a zero digit can be a multiplicative multidigital number. The number of multiplicative multidigital numbers is unlimited because the number of repunit numbers (every digit a 1) is unlimited and each repunit number is a multiplicative multidigital number.

There are numbers that are both additive *and* multiplicative multidigital numbers. One such number is 12, which is divisible by $1 + 2 = 3$ and by $1 \times 2 = 2$.

EXERCISE 97

1. Determine whether or not each of the numbers below is a multiplicative multi-digital number.

   a) 11   b) 15   c) 22   d) 24   e) 112
   f) 131   g) 144   h) 172   i) 212   j) 225

2. The numbers 12, 24, and 36 are multiples of 12. These numbers are both additive and multiplicative multidigital numbers. Are all multiples of 12 with nonzero digits both additive and multiplicative multidigital numbers? Explain your answer.

Kaprekar's Number 6174, 99 and 1089

Perhaps the most outstanding discovery of Kaprekar for which he is known outside India is the property of the remarkable number 6174. The number 6174 is called the Kaprekar number or Kaprekar's constant because he found that for any four-digit number in which the digits are not all the same, when using the algorithm described below, the number 6174 results in at most *seven* subtractions.

EXAMPLE

Start with a four-digit number, digits not all the same, say 4123.

Form the *largest* number using these digits	4321 largest
and subtract from it the *smallest* four-digit	− 1234 smallest
number made from these digits.	3087

Repeat the process with 3087.	8730 largest
	− 0378 smallest
	8352

Repeat the process with 8352.	8532 largest
	− 2358 smallest
	6174

For 4123, the Kaprekar number 6174 results after three subtractions.

EXERCISE 98

1. Use the algorithm described on the preceding page on each of the following numbers. How many subtractions are needed to reach 6174?

   a) 3261   b) 7361   c) 2378   d) 2589   e) 3535
   f) 4578   g) 1000   h) 6024   i) 2468   j) 4713

2. In each case name a four-digit number, different from the ones in the example and problem 1 that requires the given number of subtractions to reach 6174.

   a) four   b) five   c) six   d) seven

The numbers 99 and 1089 appear after a procedure similar to the one that produces the Kaprekar number, 6174. The algorithm shown in the following examples might be called a reverse-subtract, reverse-add algorithm.

EXAMPLE

Start with the two-digit number 82.	82
Reverse the digits in 82, giving 28, and	− 28
subtract the smaller number from the larger.	54
Reverse the digits in 54, giving 45, and	+ 45
add the two numbers.	99

EXAMPLE

Start with 67 and use the reverse-subtract, reverse-add algorithm.

$$
\begin{array}{r}
76 \\
-\ 67 \\
\hline
09 \\
+\ 90 \\
\hline
99
\end{array}
$$
(Include a 0 when subtraction gives a single-digit difference.)

Try to prove that the result will be 99 for any two-digit number in which both digits are not the same.

EXAMPLE

Start with the three-digit numbers 572 and 435 and use the reverse-subtract, reverse-add algorithm.

$$
\begin{array}{rr}
572 & 534 \\
-\ 275 & -\ 435 \\
\hline
297 & 099 \\
+\ 792 & +\ 990 \\
\hline
1089 & 1089
\end{array}
$$

Try to prove that the result will be 1089 for any three-digit number in which the first and third digits are different.

EXERCISE 99

1. Verify that the reverse-subtract, reverse-add algorithm yields a result of 99 for each of the two-digit numbers below.

   a) 13   b) 34   c) 52   d) 63
   e) 71   f) 89   g) 91

2. Verify that the reverse-subtract, reverse-add algorithm yields a result of 1089 for each of the three-digit numbers below.

   a) 134   b) 335   c) 677   d) 766
   e) 786   f) 814   g) 964

3. Investigate what happens with four-digit and five-digit numbers using the reverse-subtract, reverse-add algorithm.

Doubling Numbers

Doubling numbers were created by Sarah Donovan, when she was an 8th grade student at Portola Junior High School in El Cerrito, California. Her work with these numbers won her first place in the 1978 San Francisco Bay Area Science Fair.

Study the following examples carefully. They illustrate the process used for finding doubling numbers.

EXAMPLE

Is 3 a doubling number?

Start with 1 and double repeatedly until the result becomes just greater than 3.

      1
• 2
• 4 STOP

Subtract 3 from the last double, 4.
The difference is 1.

$$\begin{array}{r} 4 \\ -\ 3 \\ \hline 1 \end{array}$$

Count the *number of doublings* •, here 2. Since 2 *is* 1 *less than the given number* 3, *and* since a *difference of* 1 *is reached*, you can conclude 3 is a *doubling number.*

EXAMPLE

Is 7 a doubling number?

Start with 1 and double repeatedly until the result becomes just greater than 7.

      1
• 2
• 4
• 8 STOP

Subtract 7 from the last double, 8.   8
The difference is 1.   − 7
                                                      ———
                                                        1

Count the number of doublings •, here 3.
Since 3 is not 1 less than 7, even though a difference of 1 is reached, 7 is *not* a
doubling number.

EXAMPLE

Is 5 a doubling number?

Start with 1 and double repeatedly until the result   1
becomes just greater than 5.   • 2
                                                             • 4
                                                             • 8 STOP

Subtract 5 from the last double, 8.   8
The difference is not 1, so repeat the process   − 5
starting with the *difference*.   ———
                                                               3

Continue the doubling starting with 3, until the   3
result becomes just greater than the given   • 6 STOP
number 5.

Subtract 5 from the last double, 6.   6
Now the difference is 1.   − 5
                                                             ———
                                                             1 STOP

Count the number of doublings •, here 4. Since 4 is 1 less than the given number,
and since a difference of 1 is reached, you can conclude that 5 is a doubling number.

EXAMPLE

Is 4 a doubling number?

Start with 1 and double repeatedly until   1
the result becomes just greater than 4.   • 2
                                                     • 4
                                                     • 8 STOP

Subtract 4 from the last double, 8.   8
The difference is not 1, so repeat the process   − 4
starting with the difference.   ———
                                                        4

Continue the doubling starting with 4,
until the result becomes just greater than the
given number 4.

4
• 8 STOP

It should be clear that the pattern will *repeat* and that a difference of 1 will never be reached.
Hence 4 is not a doubling number.

EXERCISE 100

1. Which of the following are doubling numbers?

a) 8	b) 10	c) 11	d) 13	e) 16
f) 19	g) 20	h) 23	i) 25	j) 29
k) 30	l) 32	m) 37	n) 41	o) 53

2. What kind of numbers are the doubling numbers in problem 1?
3. Do you think the number of doubling numbers is unlimited? Explain your answer.

Good Numbers

Good numbers require some familiarity with unit fractions. A *unit fraction* is a fraction with numerator 1.

Fractions such as $\frac{1}{3}, \frac{1}{5}$ and $\frac{1}{13}$ are unit fractions.

An algorithm for finding good numbers is detailed in the following examples.

EXAMPLE

Is 4 a good number?

1. Write 4 as a sum of integers, not necessarily all different. Single numbers will be accepted as sums.

$$4 = 4 \qquad\qquad 4 = 3 + 1 \qquad\qquad 4 = 2 + 2$$
$$4 = 2 + 1 + 1 \qquad 4 = 1 + 1 + 1 + 1$$

Each of these equations describes a *partition* of 4.

2. Let the numbers on the right-hand side of each equation be the *denominators* of a set of unit fractions. Find the sum of each set of unit fractions.

Start with:	Write unit fractions:	Find the sum:
↓	↓	↓
$4 = 4$	$\dfrac{1}{4}$	$\dfrac{1}{4}$
$4 = 3 + 1$	$\dfrac{1}{3}, \dfrac{1}{1}$	$\dfrac{4}{3}$
$4 = 2 + 2$	$\dfrac{1}{2}, \dfrac{1}{2}$	1
$4 = 2 + 1 + 1$	$\dfrac{1}{2}, \dfrac{1}{1}, \dfrac{1}{1}$	$\dfrac{5}{2}$
$4 = 1 + 1 + 1 + 1$	$\dfrac{1}{1}, \dfrac{1}{1}, \dfrac{1}{1}, \dfrac{1}{1}$	4

3. If the *sum is 1* in at least one case, the given number is a good number. The partition $4 = 2 + 2$ yields the sum $\dfrac{1}{2} + \dfrac{1}{2} = 1$; therefore 4 is a good number.

EXAMPLE

Is 9 a good number?

It is not necessary to list all the partitions of 9. A little practice with fractions and addition leads to

$$9 = 3 + 3 + 3, \quad \text{and} \quad \frac{1}{3} + \frac{1}{3} + \frac{1}{3} = 1.$$

So 9 is a good number.

EXAMPLE

Is 3 a good number?

The partitions of 3 and the sums of the unit fractions are

$$3 = 3 \quad \text{and} \quad \frac{1}{3} \neq 1$$

$$3 = 2 + 1 \quad \text{and} \quad \frac{1}{2} + \frac{1}{1} = \frac{3}{2} \neq 1$$

$$3 = 1 + 1 + 1 \quad \text{and} \quad \frac{1}{1} + \frac{1}{1} + \frac{1}{1} = 3 \neq 1$$

Thus, 3 is not a good number.

A *good* number is a number N such that

1. $N = a_1 + a_2 + a_3 + \cdots + a_k$ and

2. $1 = \dfrac{1}{a_1} + \dfrac{1}{a_2} + \dfrac{1}{a_3} + \cdots + \dfrac{1}{a_k}$ with $a_1, a_2, a_3, \ldots, a_k$ not necessarily different integers.

It can be shown that the number of good numbers is unlimited.

EXERCISE 101

1. Determine which of the numbers from 1 through 20 is a good number. For each good number, give a partition that leads to a set of unit fractions whose sum is 1.

Nearly Good Semiperfect Numbers

Recall that a semiperfect number is a number that equals the sum of some, but not all, of its distinct proper divisors.

EXAMPLE

Show that 12 is semiperfect.
$PD_{12} = \{1, 2, 3, 4, 6\}$

$$12 = 2 + 4 + 6 \quad \text{or} \quad 12 = 1 + 2 + 3 + 6$$

So 12 is a semiperfect number.

The number 12, however, is not a good number because none of its partitions lead to a set of unit fractions whose sum is 1.

Are there semiperfect numbers that are *nearly good?*

A routine for finding nearly good semiperfect numbers is outlined in the example.

EXAMPLE

Is 12 a nearly good semiperfect number?

1. Start with the partition $12 = 2 + 4 + 6$.
2. Divide 12 by each number in the partition.

$$12 \div 2 = 6 \qquad 12 \div 4 = 3 \qquad 12 \div 6 = 2$$

3. Let the *quotients* 6, 3, and 2 be the denominators of unit fractions.
   Then compute their sum. $\dfrac{1}{6} + \dfrac{1}{3} + \dfrac{1}{2} = 1$

4. Since $12 = 2 + 4 + 6$ and $\dfrac{1}{6} + \dfrac{1}{3} + \dfrac{1}{2} = 1$, you can conclude 12 is a *nearly good semiperfect* number.

> N is a *nearly good semiperfect* number if — taking the quotients found by dividing N by each of the divisors used to show N is semiperfect, and using these quotients as denominators of unit fractions — the resulting sum of unit fractions is 1.

EXAMPLE

Is 18 a nearly good semiperfect number?

$$PD_{18} = \{1, 2, 3, 6, 9\}$$

1. 18 is semiperfect since $18 = 1 + 2 + 6 + 9$.
2. $18 \div 1 = 18, 18 \div 2 = 9, 18 \div 6 = 3, 18 \div 9 = 2$
3. $\dfrac{1}{18} + \dfrac{1}{9} + \dfrac{1}{3} + \dfrac{1}{2} = 1$
4. Since $18 = 1 + 2 + 6 + 9$ and $\dfrac{1}{18} + \dfrac{1}{9} + \dfrac{1}{3} + \dfrac{1}{2} = 1$, 18 is a nearly good semiperfect number.

EXERCISE 102

1. Which of the following semiperfect numbers are nearly good?

   a) $20 = 1 + 4 + 5 + 10$ b) $24 = 4 + 8 + 12$
   c) $24 = 1 + 2 + 3 + 4 + 6 + 8$ d) 30
   e) 40   f) 48
   g) 54   h) 60
   i) 66   j) 72

2. Do you think that all semiperfect numbers are nearly good? Explain your answer.

Powerful Numbers

The label "powerful numbers" has multiple descriptions. In 1968, the following definition was suggested by Jeremy Randle of Yorkshire, England.

Powerful numbers are integers that can be written as a sum of positive integral powers of their digits.

All the integers from 1 through 9 are powerful numbers because $1 = 1^1, 2 = 2^1, \ldots, 9 = 9^1$. Verify none of the integers from 10 through 23 are powerful numbers.

EXAMPLE

Show that 24, 43, 63, 89, and 132 are powerful numbers.

$24 = 2^3 + 4^2$ $43 = 4^2 + 3^3$ $63 = 6^2 + 3^3$
$89 = 8^1 + 9^2$ $132 = 1^1 + 3^1 + 2^7$

Some powerful numbers have more than one representation. Multiple representations occur routinely when a number has a digit 1 or 0, as any positive integral power of 1 or 0 may be used. This is also the case with numbers that have repeated digits.

EXAMPLE

Show two different representations of the powerful numbers 264 and 1323.

$$264 = 2^1 + 6^1 + 4^4 = 2^5 + 6^3 + 4^2$$
$$1323 = 1^1 + 3^4 + 2^9 + 3^6 = 1^5 + 3^6 + 2^9 + 3^4$$

EXERCISE 103

1. Name five other powerful numbers.
2. Name five powerful prime numbers.

Armstrong Numbers and Digital Invariant Numbers

In the 1960s, Michael F. Armstrong, a computer science teacher at the University of Rochester, NY gave his class an assignment based on what came to be known as Armstrong numbers, a special type of powerful numbers.

A three-digit number *abc* is an *Armstrong* number if

$$abc = a^3 + b^3 + c^3.$$

A four-digit number *abcd* is an *Armstrong* number if

$$abcd = a^4 + b^4 + c^4 + d^4.$$

A similar pattern is applied to numbers with five or more digits.

Armstrong numbers are also called *perfect digital invariant* numbers.

EXAMPLE

Show that 153 is a three-digit Armstrong number.
Because $153 = 1^3 + 5^3 + 3^3$, 153 is an Armstrong number.

Recurring digital invariant numbers are also related to powerful numbers.

The example that follows shows how to decide if a number is a third order recurring digital invariant.

EXAMPLE

Is 55 a third order recurring digital invariant number?
Start with 55, raise its digits to the third power and add.

$$5^3 + 5^3 = 125 + 125 = 250.$$

Repeat with 250. $2^3 + 5^3 + 0^3 = 8 + 125 + 0 = 133$.

Repeat with 133. $1^3 + 3^3 + 3^3 = 1 + 27 + 27 = 55$.

The chain $55 \rightarrow 250 \rightarrow 133 \rightarrow 55$ results.

The number 55 is called a *third order recurring digital invariant* number. It is a *third order* number because the sums of the *cubes* of the digits were used and it is a *recurring digital invariant* number since repeated application of the process resulted in the starting number.

Another type of number related to the powerful numbers are *amicable digital invariant* numbers.

The next example shows how to determine if a number is an amicable digital invariant of the third order.

EXAMPLE

Is 136 a third order amicable digital invariant number?

Start with 136, cube its digits and add.

$$1^3 + 3^3 + 6^3 = 1 + 27 + 216 = 244.$$

Repeat with 244. $2^3 + 4^3 + 4^3 = 8 + 64 + 64 = 136.$

The chain $136 \rightarrow 244 \rightarrow 136$ results.

The pair of numbers 136 and 244 are called *amicable digital invariant* numbers of the *third order*.

EXERCISE 104

1. Which of the following numbers are Armstrong numbers?
   a) three-digit: 370; 455; 371; 407
   b) four-digit: 1634; 2157; 8208; 9474
   c) five-digit: 12,345; 54,748; 92,727; 93,084
   d) six-digit: 315,423; 324,513; 352,612; 548,834
2. Are there any two-digit Armstrong numbers? Explain your answer.
3. Which of the following three-digit numbers are *third order* recurring digital invariant numbers?

   a) 160   b) 456   c) 578   d) 919

4. Which of the following four-digit numbers are *fourth order* recurring digital invariant numbers?

   a) 1111   b) 2178   c) 2525

5. The number 1138 is a *fourth order* recurring digital invariant number. Verify that it has a chain that consists of eight numbers.
6. 6514 is one number in a pair of amicable digital invariant numbers of the *fourth order*. Find the other number.

Narcissistic Numbers

Narcissistic numbers derive their name from Narcissus. The story is told in Greek mythology that Narcissus gazed into a pool of water and fell in love with his own image, tumbled into the water, and drowned. Subsequently, the flower that bears his name grew in his place.

In 1966, Joseph S. Madachy (1927–2014) of Ohio, prominent in recreational mathematics, defined a narcissistic number as one that can be represented by *mathematically* manipulating its digits. The following modification will be used here.

> A *narcissistic* number is a number that can be represented by taking its digits in the order *given* in combination with some mathematical operations.

EXAMPLE

Armstrong numbers are narcissistic numbers.
Recall the Armstrong number 153.
Since $153 = 1^3 + 5^3 + 3^3$, 153 is narcissistic.

EXAMPLE

Powerful numbers are narcissistic numbers.
Consider the powerful numbers 24 and 43.
Since $24 = 2^3 + 4^2$ and $43 = 4^2 + 3^3$, then 24 and 43 are narcissistic.

EXAMPLE

Show that 355, 145, and 81 are narcissistic numbers.

$$355 = (3)(5!) - 5$$
$$145 = 1! + 4! + 5!$$
$$81 = (8 + 1)^2$$

EXERCISE 105

Using some combination of the operations $+$, $-$, $\times$, $!$, $\sqrt{\ }$, $\sqrt[3]{\ }$ and raising to powers, show that the following are narcissistic numbers.

1. 24	2. 25	3. 27	4. 36
5. 39	6. 48	7. 64	8. 89

Additive Digital Root Numbers

> An *additive digital root* number of a number is one of the numbers 1, 2, 3, ... , 9. Every integer has a corresponding additive digital root number found by summing the digits of the integer, repeatedly if necessary.

EXAMPLE

Find the additive digital root number of 79.

Sum the digits of 79. $7 + 9 = 16$

Continue the process until a single digit is reached.

$$1 + 6 = 7$$

7 is the additive digital root number for 79.

Write $79 \to 16 \to 7$.

Ways other than adding digit by digit to find the additive digital root number of a larger number are given in the next example.

EXAMPLE

Find the additive digital root number of 47,365.

Break 47,365 into *any* combination of digits and proceed to find the single digit sum. Two possibilities are

- $473 + 65 = 538,$ $53 + 8 = 61,$ $6 + 1 = 7$
- $47 + 365 = 412,$ $41 + 2 = 43,$ $4 + 3 = 7$

The additive digital root number of 47,365 is 7.

Infinitely many integers have the same additive digital root number. For instance, $11, 110, 1100, 11000, \ldots$ all have an additive digital root of 2.

Additive digital root numbers provide a *negative* test for triangular numbers. It can be shown that if the additive digital root number of a given integer is 2, 4, 5, 7, or 8, the integer is *not* a triangular number. Otherwise, the integer may or may not be a triangular number.

EXAMPLE

Is 79 a triangular number?

$$79 \to 16 \to 7$$

The additive digital root number of 79 is 7. Hence 79 is *not* a triangular number.

EXAMPLE

Is 435 a triangular number?

$$435 \to 12 \to 3$$

The additive digital root number of 435 is 3, so 435 *may* be a triangular number.

Since $435 = \dfrac{29(30)}{2}$, it *is* a triangular number.

EXAMPLE

Is 57 a triangular number?

$$57 \rightarrow 12 \rightarrow 3$$

The additive digital root number of 57 is 3, so 57 *may* be a triangular number. Since $\dfrac{10(11)}{2} = 55$ and $\dfrac{11(12)}{2} = 66$, the number 57 is *not* a triangular number.

EXERCISE 106

1. Find the additive digital root numbers of each of the integers below.

   a) 188   b) 194   c) 295   d) 300
   e) 999   f) 1435   g) 34,526   h) 781,456
   i) 2,365,879   j) 46,312,345

2. List each of the additive digital root numbers from 1 through 9. Then find four integers greater than 200 that correspond to each additive digital root number.
3. Use the additive digital root number test to indicate whether each of the following integers is *not* a triangular, or is *perhaps* a triangular number.

   a) 466   b) 676   c) 742
   d) 992   e) 1176   f) 1540

4. Give an explanation that supports the statement: if the additive digital root of a given integer is 2, 4, 5, 7 or 8, the integer is *not* a triangular number.

Additive Persistence of Integers

In 1974, Harvey J. Hindin of New York, with professional interests in electrical engineering, chemistry, and mathematics, introduced the notion of additive persistence.

> The *additive persistence* of an integer is the *number* of steps required to find the additive digital root number of the given integer when adding digit by digit.

EXAMPLE

Find the additive persistence of 98.

$$98 \rightarrow 17 \rightarrow 8$$

Two steps are required to reach 8. Thus 98 has an additive persistence of 2.

EXAMPLE

Find the additive persistence of 91,919.

$$91,919 \rightarrow 29 \rightarrow 11 \rightarrow 2$$

Three steps are required to reach 2. Thus 91,919 has an additive persistence of 3.

Study the pattern in the next example. Notice that the last number in each row is the additive digital root number.

EXAMPLE

10 has an additive persistence of 1. $10 \rightarrow 1$

The sum of the digits of 19 is 10.

19 has an additive persistence of 2. $19 \rightarrow 10 \rightarrow 1$

The sum of the digits of 199 is 19.

199 has an additive persistence of 3. $199 \rightarrow 19 \rightarrow 10 \rightarrow 1$

An integer the sum of whose digits is 199 has an additive persistence of 4. Can you name one?

EXERCISE 107

1. Find the additive persistence of each of the integers below.

a) 28	b) 34	c) 37	d) 63	e) 71
f) 84	g) 109	h) 195	i) 219	j) 559
k) 667	l) 887	m) 1234	n) 8198	o) 9299

2. Study the pattern for $11 \rightarrow 2$ below, then beginning with $12 \rightarrow 3$ and with $15 \rightarrow 6$, name two numbers that have an additive persistence of 3.

   Additive persistence 1: $11 \rightarrow 2$
   Additive persistence 2: $29 \rightarrow 11 \rightarrow 2$
   Additive persistence 3: $9992 \rightarrow 29 \rightarrow 11 \rightarrow 2$

3. Name the smallest integer that has an additive persistence

   a) of 1.   b) of 2.   c) of 3.   d) of 4.

4. Name the largest integer that has an additive persistence of 1.

Multiplicative Digital Root Numbers

> A *multiplicative digital root* number of an integer is one of the numbers $0, 1, 2, 3, \ldots , 9$. Every integer has a corresponding multiplicative digital root number found by multiplying the digits of the integer, repeatedly if necessary.

EXAMPLE

Find the multiplicative digital root number of 79.

Multiply the digits of 79. Continue the process until a single digit is reached.

$7 \times 9 = 63$ $6 \times 3 = 18$ $1 \times 8 = 8$

8 is the multiplicative digital root number of 79.

Write $79 \rightarrow 63 \rightarrow 18 \rightarrow 8$.

EXAMPLE

Find the multiplicative digital root number of 109.

$$1 \times 0 \times 9 = 0$$

0 is the multiplicative digital root number of 109.

How can you tell by inspection when a given number has 0 for a multiplicative digital root number?

The multiplicative digital root number of a single digit number is that number. That is, the multiplicative digital root number of 1 is 1, 2 is 2, and so on. Infinitely many integers have the same multiplicative digital root number. For instance, 85, 109, and 425 all have a multiplicative digital root number of 0. How would you show that infinitely many integers have a multiplicative digital root number of 0?

EXERCISE 108

1. Find the multiplicative digital root number of each of the following integers.

   a) 25   b) 34   c) 49   d) 68   e) 77
   f) 839   g) 915   h) 1234   i) 2617   j) 4539

2. List each of the multiplicative digital root numbers from 0 through 9. Then find five integers greater than 200 that correspond to each multiplicative digital root number.

Multiplicative Persistence of Integers

Neil J. A. Sloane is a native of Wales, a distinguished mathematician and former researcher at AT&T Bell Laboratories in New Jersey. He has a sustained interest in number sequences and oversees the Online Encyclopedia for Integer Sequences Foundation (OEIS) that publishes an extensive database of sequences together with the related research sources. Sloane introduced the notion of multiplicative persistence in 1973.

> The *multiplicative persistence* of an integer is the *number* of steps required to find the multiplicative digital root number of the given integer.

EXAMPLE

Find the multiplicative persistence of 77.

$$77 \rightarrow 49 \rightarrow 36 \rightarrow 18 \rightarrow 8$$

Four steps are required to reach 8. So 77 has a multiplicative persistence of 4.

Study the pattern in the next example. Notice that the last number in each row is the multiplicative digital root number.

EXAMPLE

Find numbers with multiplicative persistence 1, 2, 3, and 4.
15 has a multiplicative persistence of 1.

$$15 \rightarrow 5$$

35 has a multiplicative persistence of 2.

$$35 \rightarrow 15 \rightarrow 5$$

75 has a multiplicative persistence of 3.

$$75 \rightarrow 35 \rightarrow 15 \rightarrow 5$$

355 has a multiplicative persistence of 4.

$$355 \rightarrow 75 \rightarrow 35 \rightarrow 15 \rightarrow 5$$

An integer, the product of whose digits is 355, has a multiplicative persistence of 5. Can you find one?

EXERCISE 109

1. Find the multiplicative persistence of each of the integers below.

a) 12	b) 14	c) 25	d) 26	e) 32
f) 34	g) 39	h) 40	i) 48	j) 68
k) 88	l) 97	m) 128	n) 238	o) 297

2. Study the pattern for $16 \rightarrow 6$ below, then beginning with $18 \rightarrow 8$ and with $10 \rightarrow 0$, name two numbers that have a multiplicative persistence of 3.

   Multiplicative persistence 1: $16 \rightarrow 6$
   Multiplicative persistence 2: $28 \rightarrow 16 \rightarrow 6$
   Multiplicative persistence 3: $47 \rightarrow 28 \rightarrow 16 \rightarrow 6$

3. Name the smallest integer that has a multiplicative persistence

   a) of 1.   b) of 2.   c) of 3.   d) of 4.

4. Find some numbers with a multiplicative persistence of 5. Explain your choices.
5. Name the possible multiplicative persistences of a four-digit number.

Modest and Extremely Modest Numbers

Modest numbers were created by Hans Havermann of Ontario, Canada in 1984.

> A *modest* number is a number greater than 10 that can be sectioned into two parts having the property that when the *given* number is divided by the *second* part, the remainder is the *first* part.

A given two-digit number *ab*, can be sectioned as *a*|*b*, where *a* is the first part and *b* is the second part.

A given three-digit number *abc*, can be sectioned as *a*|*bc*, where *a* is the first part and *bc* is the second part; or it can be sectioned as *ab*|*c*, where *ab* is the first part and *c* is the second part.

In division, since the remainder must be less than the divisor, *ab*|*c* is not allowable.

EXAMPLE

Find all modest numbers less than 16.

11	(sectioned as 1|1) is not modest since $11 \div 1$ has remainder 0.
12	(sectioned as 1|2) is not modest since $12 \div 2$ has remainder 0.
13	(sectioned as 1|3) is modest since $13 \div 3$ has remainder 1.
14	(sectioned as 1|4) is not modest since $14 \div 4$ has remainder 2.
15	(sectioned as 1|5) is not modest since $15 \div 5$ has remainder 0.

> If a number is modest with respect to every allowable sectioning of itself, then the number is *extremely modest*.

EXAMPLE

Decide if 1333 is extremely modest.
The sectionings of 1333 are

133|3   13|33   1|333

133|3 is not allowable.
For sectioning 13|33, $1333 \div 33$ has a remainder of 13.
For sectioning 1|333, $1333 \div 333$ has a remainder of 1.

Thus 1333 is extremely modest.

EXERCISE 110

1. Find all two-digit modest numbers.
2. a) Verify that 103 is a modest number.
    b) Describe how 206 and 309 are related to 103.
    c) Check to see if 206 and 309 are modest numbers.

3. Each number given is modest. Use the relation in problem 2 to find some more three-digit modest numbers.

   a) 133   b) 203   c) 211   d) 433

4. Describe the allowable ways in which numbers in these forms can be sectioned.

   a) *abcd*   b) *abcde*

5. Determine which of the following four-digit numbers are modest.

   a) 2333   b) 2424   c) 2666   d) 2817

6. Is every four-digit number of the forms given below where *a* is one of the digits 1, 2, ... , 9, extremely modest? Explain your answer.

   a) *a*333   b) *a*666   c) *a*999

Visible Factor Numbers

Visible factor numbers were the creation of J. A. Lindon of Surrey, England in 1968. A *modification* of this kind of number is described below.

Visible factor numbers are numbers that are divisible by each of the nonzero digits that form the number.

EXAMPLE

Are 10, 12, and 13 visible factor numbers?

10 is a visible factor number because 10 is divisible by 1.
12 is a visible factor number because 12 is divisible by 1 and 2.
13 is not a visible factor number because 13 is not divisible by 3.

EXERCISE 111

1. Which of the following numbers are visible factor numbers?

   a) 15   b) 18   c) 22   d) 24   e) 28
   f) 36   g) 48   h) 56   i) 62   j) 75
   k) 135   l) 248   m) 486   n) 952

2. Which of the numbers below are divisible by their *prime* digits?

   a) 147   b) 153   c) 165   d) 182   e) 267
   f) 285   g) 375   h) 571   i) 593   j) 728

Nude Numbers

Yoshinao Katagiri of Chigasaki, Japan developed nude numbers in 1982. A nude number exposes many of its divisors in its representation.

> A number N is *nude* if it is divisible by all nonzero digits of N. If N has more than two digits then it must be divisible by at least one number, other than N itself that is formed of two or more digits of N.

Every one-digit integer is nude. If 0 occurs as a digit ignore it. Thus 30 is a nude number — as it is divisible by 3.

EXAMPLE

Find all two-digit nude numbers greater than 30 and less than 40.
The numbers 31, 32, 34, 35, 37, 38, and 39 are not nude.
The number 33 is nude since $33 \div 3 = 11$.
The number 36 is nude, since $36 \div 3 = 12$, $36 \div 6 = 6$.

EXAMPLE

What divisors of the nude number 120 are exposed in the representation of the number?
120 has 16 divisors and 6 of them are exposed. The divisors that are formed by the digits 0, 1, and 2 are exposed. They are 1, 2, 10, 12, 20, and 120.

EXERCISE 112

1. Find all two-digit nude numbers.
2. Is it true that if ab is a nude number then cab may be a nude number when c is not zero? Explain your answer.
3. Is it true that if ab is any nude number then $abab$ is a nude number? Explain your answer.
4. Which permutations of the numbers below are three-digit nude numbers?

   a) 213   b) 639   c) 428   d) 537

5. Construct some four-digit nude numbers by placing 2 two-digit multiples of 12 side by side.
6. Name some nude numbers that have 2 or more two-digit divisors.
7. Name some nude numbers that have some two- and three-digit divisors.
8. Show that 27,216 is nude and that it is divisible by at least 8 visible factors.
9. Compare visible factor numbers and nude numbers.

Chapter 7

More Patterns and Other Interesting Numbers

[For a large part] of my life all I wanted to work in was number theory ... my dream subject.

Olga Taussky-Todd (1906–1995)

Number enthusiasts continue to make new discoveries about the integers. This collection is a sampling of some of these results. It is intended to serve as a catalyst and encouragement for readers to undertake their own investigations.

More Sums of Consecutive Integers

The methods given previously for finding the sums of consecutive integers starting with 1 also work for sums of consecutive integers that do not begin with 1.

EXAMPLE

Odd number of terms in the series

$$3 + 4 + 5 = 3 \times 4 = 12$$

$$4 + 5 + 6 + 7 + 8 = 5 \times 6 = 30$$

$$5 + 6 + 7 + 8 + 9 + 10 + 11 = 7 \times 8 = 56$$

That is, the sum equals the number of terms in the series times the middle term of the series.

EXAMPLE

Even number of terms in the series

$$2 + 3 + 4 + 5 = 2 \times (2 + 5) = 14$$

$$5 + 6 + 7 + 8 + 9 + 10 = 3 \times (5 + 10) = 45$$

$$8 + 9 + 10 + 11 + 12 + 13 + 14 + 15 = 4 \times (8 + 15) = 92$$

That is, the sum equals the number of pairs of terms in the series times the sum of the first and last term of the series.

EXAMPLE

A way to sum four consecutive integers

$$\underline{2+3}+\underline{4+5}$$

Find the products 2×3 and 4×5.
Take the difference of the products $(4 \times 5) - (2 \times 3) = 14$.
14 *is* the sum of the series.

Can you explain why this procedure works?

EXERCISE 113

1. Find a way to tell how many terms are in each of the following series by using only the first and last terms.
   Then compute the sum.

   a) $9 + 10 + 11 + 12$ b) $12 + 13 + 14 + \cdots + 16$
   c) $4 + 5 + 6 + \cdots + 10$ d) $6 + 7 + 8 + \cdots + 11$
   e) $7 + 8 + 9 + \cdots + 15$ f) $15 + 16 + 17 + \cdots + 22$
   g) $18 + 19 + 20 + \cdots + 27$ h) $20 + 21 + 22 + \cdots + 30$
   i) $25 + 26 + 27 + \cdots + 36$ j) $30 + 31 + 32 + \cdots + 50$

2. Find the sum of each of the following pairs of consecutive numbers.

   a) $10 + 11$ b) $18 + 19$ c) $29 + 30$
   d) $41 + 42$ e) $100 + 101$ f) $154 + 155$
   g) $1001 + 1002$ h) $4509 + 4510$ i) $6789 + 6790$

3. Is the sum of *any* pair of consecutive integers an odd integer? Give a reason.
4. Write each of the following odd numbers as the sum of two consecutive integers.

   a) 11   b) 19   c) 37   d) 49
   e) 53   f) 65   g) 81   h) 109

5. Write the odd number $2n - 1$ as the sum of two consecutive integers.
6. Find the sum of each of the following triples of consecutive numbers.

   a) $10 + 11 + 12$ b) $14 + 15 + 16$ c) $20 + 21 + 22$
   d) $42 + 43 + 44$ e) $91 + 92 + 93$ f) $100 + 101 + 102$
   g) $203 + 204 + 205$ h) $502 + 503 + 504$ i) $989 + 990 + 991$

7. In problem 6, the sums can be found by addition or by a short cut. Describe a short cut.
8. Is the sum of *any* three consecutive integers always a multiple of 3? Give a reason.

Now consider the reverse of problem 6 in **EXERCISE 113**. That is, given a multiple of 3, how do you find the three consecutive numbers whose sum is the multiple of 3?

EXAMPLE

Find three consecutive integers whose sum is 66.

Divide 66 by 3. $66 \div 3 = 22$

Using 22, subtract and add 1 to get the other two numbers.

$$66 = 21 + 22 + 23$$

Note that 66 is an even number as well as a multiple of 3. We can also write 66 as the sum of three consecutive *even* integers. After dividing 66 by 3, simply subtract and add 2 to the result.

$$66 = 20 + 22 + 24$$

EXERCISE 114

1. Write each of the integers below as the sum of three consecutive integers.

   a) 30   b) 42   c) 60   d) 87   e) 93
   f) 99   g) 111   h) 216   i) 333

2. Write each of the following multiples of 3 as the sum of three consecutive *even* or three consecutive *odd* integers.

   a) 21   b) 54   c) 81   d) 93   e) 102
   f) 108   g) 114   h) 195   i) 243

3. Find the sum of each quadruple of consecutive numbers.

   a) $5 + 6 + 7 + 8$ b) $10 + 11 + 12 + 13$
   c) $12 + 13 + 14 + 15$ d) $18 + 19 + 20 + 21$
   e) $22 + 23 + 24 + 25$ f) $30 + 31 + 32 + 33$
   g) $40 + 41 + 42 + 43$ h) $51 + 52 + 53 + 54$

4. In problem 3, the sums can be found by addition or by a short cut. Can you describe a shortcut?

5. Is the sum of four consecutive integers always the double of an odd number? Give a reason.

6. Write each of the following doubles of an odd number as a sum of four consecutive integers.

   a) 34   b) 42   c) 50   d) 62   e) 78
   f) 102   g) 126   h) 142   i) 150

7. Find the sum of each quintuple of consecutive numbers.

   a) $4 + 5 + 6 + 7 + 8$ b) $5 + 6 + 7 + 8 + 9$
   c) $6 + 7 + 8 + 9 + 10$ d) $7 + 8 + 9 + 10 + 11$
   e) $8 + 9 + 10 + 11 + 12$ f) $9 + 10 + 11 + 12 + 13$
   g) $23 + 24 + 25 + 26 + 27$ h) $54 + 55 + 56 + 57 + 58$

8. In problem 7, the sums can be found by addition or by a short cut. Can you describe a shortcut?
9. Is the sum of five consecutive integers always a multiple of 5? Give a reason.
10. Write each of the following multiples of 5 as a sum of five consecutive integers.

   a) 75   b) 95   c) 105   d) 175
   e) 225   f) 255   g) 300   h) 350

11. Which one(s) of these powers of two — 1, 2, 4, 8, 16, and 32 — can be written as a sum of consecutive integers?
12. Write each integer below as the sum of consecutive integers in as many ways as possible.

   a) 12   b) 15   c) 18   d) 27   e) 30   f) 35
   g) 37   h) 45   i) 54   j) 55   k) 60   l) 63

13. **Challenge** For any integer, find the least number of consecutive integers whose sum equals the given integer.

    Each triangular number is the sum of consecutive integers starting with 1.

Triangular number	Sum
1st	$1 = 1$
2nd	$3 = 1 + 2$
3rd	$6 = 1 + 2 + 3$
4th	$10 = 1 + 2 + 3 + 4$
5th	$15 = 1 + 2 + 3 + 4 + 5$

From problem 12 in **EXERCISE 114** the triangular number 15 can also be written as $15 = 7 + 8$ and $15 = 4 + 5 + 6$. Thus, some triangular numbers can be written as a sum of consecutive integers that do not begin with 1.

In some problems, it helps to know the possible endings of the sums of two, three, or more consecutive integers.

EXAMPLE

What are all the possible endings for the sums of any *two* consecutive integers? Compute all sums of two of the digits 0 through 9 taken consecutively.

$$0 + 1 = 1 \qquad 1 + 2 = 3 \qquad 2 + 3 = 5 \qquad 3 + 4 = 7 \qquad 4 + 5 = 9$$
$$5 + 6 = 1\underline{1} \qquad 6 + 7 = 1\underline{3} \qquad 7 + 8 = 1\underline{5} \qquad 8 + 9 = 1\underline{7} \qquad 9 + 0 = \underline{9}$$

Thus, the possible endings for the sums of any *two* consecutive integers are the odd numbers 1, 3, 5, 7, and 9.

EXERCISE 115

1. If possible, write each triangular number below as a sum of consecutive integers that *do not* begin with 1.

   a) 21   b) 28   c) 36   d) 45

   e) 55   f) 66   g) 78   h) 91

2. Find all the possible endings for the sums of the number of consecutive integers given below.

   a) 3   b) 4   c) 5   d) 6

   e) 7   f) 8   g) 9   h) 10

Product Patterns for Consecutive Integers

The square of the middle integer of three consecutive integers appears to be the product of the other two integers plus 1.

EXAMPLE

For 1, 2, 3: $2^2 = 4$ and $1(3) + 1 = 4$.
For 2, 3, 4: $3^2 = 9$ and $2(4) + 1 = 9$.

Any three consecutive integers contain an integer that is a multiple of 3. For instance we have (1, 2, $\underline{3}$), or (8, $\underline{9}$, 10), or ($\underline{12}$, 13, 14). The next example illustrates that the product of any three consecutive integers appears to be divisible by 6.

EXAMPLE

For (1, 2, 3): $1 \times 2 \times 3 = 6$ and $6 \div 6 = 1$.
For (2, 3, 4): $2 \times 3 \times 4 = 24$ and $24 \div 6 = 4$.
For (7, 8, 9): $7 \times 8 \times 9 = 504$ and $504 \div 6 = 84$.

EXERCISE 116

1. Show that for each of the following triples of consecutive integers, the square of the middle integer equals the product of the other two integers plus 1.

   a) (9, 10, 11)   b) (10, 11, 12)   c) (13, 14, 15)

   d) (20, 21, 22)   e) (25, 26, 27)   f) (30, 31, 32)

2. Why will the pattern in problem 1 hold for *any* three consecutive integers?

3. Show that for each of the following triples of consecutive numbers, the product is divisible by 6.

a) (20, 21, 22)   b) (25, 26, 27)   c) (30, 31, 32)
d) (35, 36, 37)   e) (41, 42, 43)   f) (55, 56, 57)

4. Will the product of *any* three consecutive integers be divisible by 6? Explain your choice.

The next example illustrates that the product of four consecutive integers plus 1 appears to be a square number.

EXAMPLE

For (1, 2, 3, 4): $1 \times 2 \times 3 \times 4 + 1 = 25 = 5^2$.
For (2, 3, 4, 5): $2 \times 3 \times 4 \times 5 + 1 = 121 = 11^2$.

In the following example, there is evidence that the product of two consecutive integers less the first integer appears to equal a square number.

EXAMPLE

For (1, 2): $1(2) - 1 = 2 - 1 = 1^2$.
For (2, 3): $2(3) - 2 = 6 - 2 = 2^2$.
For (3, 4): $3(4) - 3 = 12 - 3 = 3^2$

EXERCISE 117

1. Show that for each of the following quadruples of consecutive integers, the product plus 1 is a square number.

a) (3, 4, 5, 6)   b) (4, 5, 6, 7)
c) (5, 6, 7, 8)   d) (6, 7, 8, 9)
e) (7, 8, 9, 10)   f) (8, 9, 10, 11)
g) (9, 10, 11, 12)   h) (10, 11, 12, 13)
i) (11, 12, 13, 14)   j) (12, 13, 14, 15)

2. Will the product of *any* four consecutive integers plus 1 be a square number? Explain your choice.
3. Show that for each of the pairs of consecutive integers below, the product less the first integer is a square number.

a) (4, 5)   b) (5, 6)   c) (6, 7)   d) (7, 8)
e) (8, 9)   f) (12, 13)   g) (15, 16)   h) (19, 20)
i) (24, 25)   j) (30, 31)

4. Will the product of *any* two consecutive integers less the first integer equal a square number? Explain your choice.

Consecutive Integer Divisors

Suppose you are given the digits 0, 1, 2, 3, 4, 5, 6, 7, 8. 9. Choose three digits. You may select three different digits, or you may choose the same digit twice. Using the three chosen digits, form all possible three-digit numbers. From this group make a list of three-digit numbers, not necessarily all different, so the first number is divisible by 1, the second number is divisible by 2, the third number is divisible by 3, and so on. The list should be as long as possible.

EXAMPLE

Using the digits 0, 1, and 5, create a list of three-digit numbers divisible by the most consecutive integers.

The three-digit numbers are 015, 051, 150, 105, 510, and 501.

One possible list	Another possible list
015 ← divisible by 1	105 ← divisible by 1
150 ← divisible by 2	510 ← divisible by 2
051 ← divisible by 3	105 ← divisible by 3

Still other lists of three numbers could be formed. It is not possible to make a longer list since none of the six numbers is divisible by 4.

EXAMPLE

Using the digits 5, 2, and 2, create a list of three-digit numbers divisible by the most consecutive integers.

The three-digit numbers are 522, 252, and 225.

522, 252, or 225	←	divisible by 1
522 or 252	←	divisible by 2
522, 252, or 225	←	divisible by 3
252	←	divisible by 4
225	←	divisible by 5
522 or 252	←	divisible by 6
252	←	divisible by 7

None of the numbers is divisible by 8. Thus the longest list is one that has seven numbers, since the consecutive numbers from 1 through 7 divide at least one of the three-digit numbers.

EXERCISE 118

1. Using the digits given below, create a list of three-digit numbers divisible by the most consecutive integers.

   a) 1, 2, 3   b) 2, 4, 6   c) 1, 3, 5
   d) 0, 3, 6   e) 0, 1, 2   f) 3, 6, 9

2. Look for patterns in the following table. Then extend the table to the consecutive integer divisors 1, 2, ... , 10.

Consecutive integer divisors	Smallest integer divisible exactly by the consecutive integer divisors
1	1
1, 2	2
1, 2, 3	6
1, 2, 3, 4	12

Consecutive Number Sums and Square Numbers

There are a number of ways that consecutive number sums are related to odd and even square numbers. Three of these relationships are explored in the examples and exercises below.

EXAMPLE

$$1 + (3) = 4 = 2^2$$
$$1 + (4 + 5 + 6) = 16 = 4^2$$
$$1 + (5 + 6 + 7 + 8 + 9) = 36 = 6^2$$
$$1 + (6 + 7 + 8 + 9 + 10 + 11 + 12) = 64 = 8^2$$

EXAMPLE

$$1 = 1 = 1^2$$
$$2 + 3 + 4 = 9 = 3^2$$
$$3 + 4 + 5 + 6 + 7 = 25 = 5^2$$
$$4 + 5 + 6 + 7 + 8 + 9 + 10 = 49 = 7^2$$

EXAMPLE

$$1 = 1 = 1^2, \quad \text{a sum of 1 odd integer}$$
$$1 + 3 = 4 = 2^2, \quad \text{a sum of 2 odd integers}$$
$$1 + 3 + 5 = 9 = 3^2, \quad \text{a sum of 3 odd integers}$$
$$1 + 3 + 5 + 7 = 16 = 4^2, \quad \text{a sum of 4 odd integers}$$

EXERCISE 119

1. Find each of the sums below and express it as a square number.

   a) $1 + (7 + 8 + 9 + \cdots + 15)$
   b) $1 + (8 + 9 + 10 + \cdots + 18)$
   c) $1 + (9 + 10 + 11 + \cdots + 21)$
   d) $1 + (10 + 11 + 12 + \cdots + 24)$
   e) $1 + (11 + 12 + 13 + \cdots + 27)$
   f) $1 + (12 + 13 + 14 + \cdots + 30)$

2. In a sentence describe the pattern in problem 1.
3. Find each of the sums below and express it as a square number.

   a) $5 + 6 + 7 + \cdots + 13$ b) $6 + 7 + 8 + \cdots + 16$
   c) $7 + 8 + 9 + \cdots + 19$ d) $8 + 9 + 10 + \cdots + 22$
   e) $9 + 10 + 11 + \cdots + 25$ f) $10 + 11 + 12 + \cdots + 28$

4. In a sentence describe the pattern in problem 3.
5. For each of the following sums of consecutive odd numbers, give the number of consecutive odd numbers in the sum, and then express the sum as a square number.

   a) $1 + 3 + 5 + \cdots + 9$ b) $1 + 3 + 5 + \cdots + 11$
   c) $1 + 3 + 5 + \cdots + 13$ d) $1 + 3 + 5 + \cdots + 15$
   e) $1 + 3 + 5 + \cdots + 17$ f) $1 + 3 + 5 + \cdots + 19$

6. In a sentence describe the pattern in problem 5.
7. Notice that

$$4 - (1 + 3) = 0 \quad \text{and} \quad \sqrt{4} = 2 \quad \text{with } 2^2 = 4$$
$$9 - (1 + 3 + 5) = 0 \quad \text{and} \quad \sqrt{9} = 3 \quad \text{with } 3^2 = 9$$

   Continue this pattern for the numbers below

   a) 16   b) 36   c) 81   d) 100   e) 169

8. Describe the pattern in problem 7 for finding a square root.

Factorial Numbers, Applications and Extensions

The expression $n!$ first appeared in connection with additive multidigital numbers on p. 156.

$n!$, read *n factorial*, indicates the product of the consecutive integers from 1 through n.
0! is defined to equal 1.

EXAMPLE

Find the value of 1!, 2!, 3!, and 4!

$$1! = 1$$
$$2! = 1 \times 2 = 2$$
$$3! = 1 \times 2 \times 3 = 6$$
$$4! = 1 \times 2 \times 3 \times 4 = 24$$

The numbers 1, 2, 6, and 24 are called *factorial numbers* because $1 = 1!$, $2 = 2!$, $6 = 3!$, and $24 = 4!$

It is easier to simplify expressions having factorial symbols by dividing out common factors *before* multiplication.

EXAMPLE

Simplify the expressions $\dfrac{5!}{3!}$ and $\dfrac{5!}{2!3!}$.

$$\frac{5!}{3!} = \frac{1 \times 2 \times 3 \times 4 \times 5}{1 \times 2 \times 3} = 20$$
$$\frac{5!}{2!3!} = \frac{1 \times 2 \times 3 \times 4 \times 5}{1 \times 2 \times 1 \times 2 \times 3} = 10$$

EXERCISE 120

1. Find the value of the factorial numbers from 5! through 11!
2. Simplify each of the following expressions.

   a) $\dfrac{6!}{3!}$ b) $\dfrac{10!}{7!}$ c) $\dfrac{8!}{6!}$

   d) $\dfrac{(8-4)!}{4!}$ e) $\dfrac{(7-2)!}{3!}$ f) $\dfrac{(11-5)!}{5!}$

   g) $\dfrac{6!}{3!3!}$ h) $\dfrac{9!}{4!5!}$ i) $\dfrac{7!}{3!4!}$

3. There is a way to check the equations below *without* performing the calculations. Can you find it?

   a) $(6!)(7!) = 10!$ b) $(3!)(5!) = 6!$
   c) $(3!)(5!)(7!) = 10!$ d) $(4!)(23!) = 24!$

4. Use the procedure in problem 3 to simplify $(2!)(4!)(47!)$.
5. Recall that the numbers on a particular diagonal of Pascal's triangle are triangular numbers. These numbers as well as other numbers in Pascal's triangle can be written using factorials. Look for a pattern in the table on p. 187. Then extend the table through the 10th triangular number.

Triangular number	Factorial form
1st	$\dfrac{2!}{2!(2-2)!} = 1$
2nd	$\dfrac{3!}{2!(3-2)!} = 3$
3rd	$\dfrac{4!}{2!(4-2)!} =$
4th	$\dfrac{5!}{2!(5-2)!} =$

Numbers of the form $\dfrac{n!}{r!(n-r)!}$ are called *binomial coefficient numbers*. The case $r = 2$, reduces to the triangular number formula.

Factorial numbers can be used to simplify counting procedures.

EXAMPLE

How many ways can you arrange one book on a shelf? Two different books? Three different books?

a) There is exactly 1 way to place book *A* on the shelf.

b) Two different books *A* and *B* can be arranged on the shelf in the order *AB* or *BA*. There are exactly 2 ways of arranging the two books on the shelf.

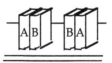

c) Three different books *A*, *B*, and *C* can be arranged on the shelf in the order *ABC*, *ACB*, *BAC*, *BCA*, *CAB*, or *CBA*. There are exactly 6 ways of arranging the three books on the shelf.

EXAMPLE

Write a string of consecutive composite numbers that contains 1 number, 2 numbers, 3 numbers.

$2! + 2 = 4$ 1 composite number

$3! + 2 = 8$
$3! + 3 = 9$ a string of 2 consecutive composite numbers

$4! + 2 = 26$
$4! + 3 = 27$
$4! + 4 = 28$ a string of 3 consecutive composite numbers

EXERCISE 121

1. Given four different books A, B, C, and D. List the different ways they can be arranged on a shelf four at a time.

2. If you used 16 different quilt squares to make a 4×4 quilt block, how many different quilt blocks could you make?

3. Compute the value of each of the expressions below. Decide if they make a string of four consecutive composite numbers.

   a) $5! + 2$ b) $5! + 3$ c) $5! + 4$ d) $5! + 5$

4. Compute the value of each of the expressions below. Decide if they make a string of five consecutive composite numbers.

   a) $6! + 2$ b) $6! + 3$ c) $6! + 4$ d) $6! + 5$ e) $6! + 6$

5. How many consecutive composite numbers do you get in each of the following collections?

   a) $7! + 2, 7! + 3, \ldots, 7! + 7$
   b) $10! + 2, 10! + 3, \ldots, 10! + 10$
   c) $100! + 2, 100! + 3, \ldots, 100! + 100$

6. Give some advantages and disadvantages of the process shown in problems 3–5. Is there any limit to the number of consecutive composite numbers you can find with this procedure? Explain your answer.

Try the **Investigations: Dealing with Digits-Base Ten** and **Factorial Finishes.**

Factorial Sum Numbers and Subfactorial Numbers

A *factorial sum* number is an integer that equals the sum of the factorials of its digits.

EXAMPLE

Is the number 145 a factorial sum number?

$1! + 4! + 5! = 1 + 24 + 120 = 145$

Therefore, 145 is a factorial sum number.

EXAMPLE

Is the number 255 a factorial sum number?

$2! + 5! + 5! = 2 + 120 + 120 = 242$

Therefore, 255 is not a factorial sum number.

Subfactorial n is defined as

$$!n = n! \left[1 - \frac{1}{1!} + \frac{1}{2!} - \frac{1}{3!} + \cdots + (-1)^n \left(\frac{1}{n!} \right) \right]$$

EXAMPLE

Since $!0 = 0![(-1)^0 \bullet (\frac{1}{0!})] = 1 \bullet 1 = 1$, 1 is a subfactorial number.

EXERCISE 122

1. Determine if any single-digit integers are factorial sum numbers.
2. Which one(s) of the following numbers is/are factorial sum numbers?

   a) 40585   b) 1266

3. Find all single-digit subfactorial numbers.
4. Compute $n!+!n$ for $n \leq 8$. What patterns do you observe?

Hailstone and Ulam Numbers; The Collatz and Ulam Conjecture

Hailstone numbers are associated with Lothar Collatz, (1910–1990) a distinguished German mathematician, Stanislaw Ulam, and many others. In 1937, Lothar Collatz conjectured that every number is a hailstone number. Just as hailstones bounce up and down in the clouds before plunging to the ground, hailstone numbers produce a sequence of numbers that rise and fall in size and eventually reach 1.

Hailstone numbers are integers that yield a sequence of integers that eventually terminates in 1 according to the following rules.

Rule 1. If the number is odd, multiply it by 3 and add 1.

Rule 2. If the number is even, divide it by 2.

EXAMPLE

Is 1 a hailstone number?

Since 1 is odd, use rule 1 to get $3(1) + 1 = 4$.
Since 4 is even, use rule 2 to get $4 \div 2 = 2$.
Since 2 is even, use rule 2 to get $2 \div 2 = 1$.
So 1 is a hailstone number and it took 3 steps to reach 1.

EXAMPLE

Is 6 a hailstone number?

Since 6 is even, use rule 2 to get 3.
Since 3 is odd, use rule 1 to get 10.
Since 10 is even, use rule 2 to get 5.
Since 5 is odd, use rule 1 to get 16.
Since 16 is even, use rule 2 to get 8.
Since 8 is even, use rule 2 to get 4.
Since 4 is even, use rule 2 to get 2.
Since 2 is even, use rule 2 to get 1.
So 6 is a hailstone number and it took 8 steps to reach 1. The peak value reached in this process is 16. Observe the powers of two in the sequence.

The conjecture has yet to be proved although it is a fact that all numbers up to at least 5×2^{60}, which is approximately 5.764×10^{18}, have been tested and every one reaches 1. It is quite probable that there are an unlimited number of hailstone numbers.

Ulam while pursuing a proof of the conjecture looked at a variation of the hailstone number rules.

Ulam numbers are integers that yield a sequence of integers that eventually end with a quotient of 1 according to the following rules.
Rule 1. If the number is odd, multiply it by 3, add 1, and divide the sum by 2.
Rule 2. If the number is even, divide it by 2.

EXAMPLE

Is 1 an Ulam number?

1 is odd, so use rule 1 to get $\dfrac{((3 \times 1) + 1)}{2} = 2$, and continue.

Since 2 is even, use rule 2 to get $\dfrac{2}{2} = 1$, and stop.

So 1 is an Ulam number, and two divisions were required to reach 1.

EXAMPLE

Is 2 an Ulam number?

Since 2 is even, use rule 2 to get $\frac{2}{2} = 1$, and stop.

So 2 is an Ulam number, and one division was required to reach 1.

EXAMPLE

Is 6 an Ulam number?

Since 6 is even, use rule 2 to get 3.
Since 3 is odd, use rule 1 to get 5.
Since 5 is odd, use rule 1 to get 8.
Since 8 is even, use rule 2 to get 4.
Since 4 is even, use rule 2 to get 2.
Since 2 is even, use rule 2 to get 1, and stop.
So 6 is an Ulam number, and six divisions are required to reach 1.

An intriguing part of this research is studying the number of steps required to reach 1 and examining the peak values as well as other number patterns that emerge in the sequences.

EXERCISE 123

1. Verify that each of the integers from 3 through 14 is an Ulam number. In each case, give the number of divisions required to reach 1.
2. Verify that each of the integers from 1 through 25 is a hailstone number. In each case give the number of steps needed to reach 1 and name the peak value reached.

Palindromic Numbers

A *palindromic* word, phrase, sentence, paragraph, or poem is a word, phrase, sentence, paragraph, or poem that is the same when read backward or forward. In mathematics, numbers that read the same backward or forward are called *palindromic numbers*.

EXAMPLE

Name some palindromic words.

DAD   MOM   LEVEL   RADAR

EXAMPLE

Name some palindromic phrases.

A TOYOTA

NEVER ODD OR EVEN

TOO HOT TO HOOT

EXAMPLE

Name some palindromic sentences.

DON'T NOD.

RISE TO VOTE, SIR.

MADAM IN EDEN, I'M ADAM.

Create some palindromic words, phrases, and sentences of your own.

EXAMPLE

Give some palindromic numbers.

1   22   414   2332   52325   123321

The following sequence of palindromic numbers: 1, 11, 121, 1221, 12221, ... is unlimited. Of the first 100,000 numbers 1% are palindromes. In fact, the number of palindromic numbers is unlimited.

EXERCISE 124

1. Make up at least four palindromic numbers with

   a) two digits   b) three digits   c) four digits
   d) five digits   e) six digits

2. Each single-digit number is a palindromic number. There are 18 palindromic numbers in the range from 1 through 100. Find the number of palindromic numbers with

   a) three digits   b) four digits   c) five digits

3. Give the first ten terms of a sequence of palindromic numbers that begins

   a) 2, 22, 222, ...   b) 4, 44, 464, ...   c) 7, 77, 707, ...

4. A palindromic decomposition of a number is a sum of two or more terms that reads the same forward and backward.

   $1 + 3 + 1$ is a palindromic decomposition of 5.
   $3 + 1 + 1$ is not a palindromic decomposition of 5.

   Find all palindromic decompositions of the following:

   a) 5   b) 6   c) 7   d) 8

Creating Palindromic Numbers

The following reverse-add algorithm usually leads to some palindromic numbers.

EXAMPLE

Begin with 12.	12
Reverse the digits in 12 and add.	+ 21
The sum, after one reversal, is a palindromic number.	33

EXAMPLE

Begin with 37.	37
Reverse the digits in 37 and add.	+ 73
110 is not a palindromic number.	110
Reverse the digits in 110 and add.	+ 011
The sum, after two reversals, is a palindromic number.	121

EXAMPLE

Begin with 59.	59
Reverse the digits in 59 and add.	+ 95
154 is not a palindromic number.	154
Reverse the digits in 154 and add.	+ 451
605 is not a palindromic number.	605
Reverse the digits in 605 and add.	+ 506
The sum, after three reversals, is a palindromic number.	1111

Although the reverse-add algorithm when applied to all two-digit numbers results in a palindrome, apparently this is not the case for some three-digit and larger numbers. The smallest of these numbers is 196. No palindromic sum has occurred in well over one billion iterations of the algorithm. All together, 12 other three-digit numbers appear to be like 196 and do not yield a palindromic sum.

Jason Doucette of Yarmouth, Nova Scotia, has produced considerable research on palindromes. He reported that about 80% of all numbers less than 10,000 reach a palindrome in 4 or less reversals and that about 90% of all numbers in that range reach a palindrome in 7 or less reversals.

The decimals 9.9, 12.21, 123.321, and 235.532 are the same when read backward or forward, and are called *palindromic decimal* numbers. The reverse-add algorithm also leads to some palindromic decimals. Note the placement of the decimal points in the following examples.

EXAMPLE

Begin with 1.8.	1.8
Reverse the digits in 1.8 and add.	+ 8.1
The sum, after one reversal, is a palindromic decimal.	9.9

EXAMPLE

Begin with 19.6.	19.6
Reverse the digits in 19.6 and add.	+ 6.91
26.51 is not a palindromic decimal.	26.51
Reverse the digits in 26.51 and add.	+ 15.62
42.13 is not a palindromic decimal.	42.13
Reverse the digits in 42.13 and add.	+ 31.24
The sum, after three reversals, is a palindromic decimal.	73.37

EXERCISE 125

1. Use the reverse-add algorithm to create a palindromic number from each of the starting numbers below.
   Indicate the number of reversals required to reach the palindromic number.

a) 14	b) 19	c) 28	d) 37	e) 68
f) 69	g) 77	h) 78	i) 79	j) 86
k) 87	l) 88	m) 89	n) 167	o) 193
p) 292	q) 490	r) 639	s) 837	t) 967

2. Make up four palindromic decimal numbers with

   a) two digits   b) four digits   c) six digits

3. Use the reverse-add algorithm to create a palindromic decimal number from each of the starting decimals below. Indicate the number of reversals required to reach the palindromic decimal number.

a) 1.2	b) 3.8	c) 5.7	d) 7.98	e) 16.8
f) 4.6	g) 6.57	h) 9.31	i) 39.8	j) 50.9

4. **Challenge** Group all two-digit numbers according to the number of reversals required to reach a palindromic sum in the reverse-add algorithm. Name the different palindromic numbers that appear as the final sum.

Palindromic Number Words and Curiosities

The activities outlined for happy numbers, on p. 96, can also be applied to palindromic numbers. For example, assign the numbers 1, 2, 3, . . . , 26 to the letters of the alphabet as shown.

A	B	C	D	E	F	G	H	I	J	K	L	M
1	2	3	4	5	6	7	8	9	10	11	12	13

N	O	P	Q	R	S	T	U	V	W	X	Y	Z
14	15	16	17	18	19	20	21	22	23	24	25	26

EXAMPLE

Is TINA a palindromic number name?

T   I   N   A
20 + 9 + 14 + 1 = 44

The number 44 is a palindromic number and so TINA is a palindromic number name.

EXAMPLE

Is JANUARY a palindromic number month?

J   A   N   U   A   R   Y
10 + 1 + 14 + 21 + 1 + 18 + 25 = 90

Since 90 is not a palindromic number, January is not a palindromic number month.

EXERCISE 126

1. Which of the following are palindromic number names?

a) LULU   b) LEE   c) DOUG   d) TODD
e) FAITH   f) KOFI   g) MOLLY   h) INEZ
i) AIKA   j) KAROL   k) GURI   l) KAREN

2. Check to see if your first name, middle name, last name, and complete name are palindromic number names.
3. Which months of the year are palindromic number months?
4. Which days of the week are palindromic number days?
5. Name the next five years that will be palindromic number years.
6. Compute the products to find some palindromic number curiosities.

a) Product pattern 1:   3(37)
303(37)
30303(37)
3030303(37)
303030303(37)

b) Product pattern 2:   73(152207)
                                 7373(152207)
                                 737373(152207)
                                 73737373 (152207)

c) Product pattern 3:   41(271)
                                 4141(271)
                                 414141(271)
                                 41414141(271)
                                 4141414141(271)

Palindromic Numbers and Figurate Numbers

Are any figurate numbers also palindromic numbers?

Some triangular numbers are palindromic. For instance, 1, 3, 6, and 55 are both triangular and palindromic. But there are only 28 numbers that are both triangular and palindromic less than 10^{10}!

Some square numbers are also palindromic numbers. For instance, $26^2 = 676$ is palindromic, even though 26 is not palindromic. In addition, there are palindromic numbers. whose squares are palindromic, such as the numbers 1, 2, 3, and 11 and their squares 1, 4, 9, and 121.

The cubes of some palindromic numbers are also palindromic. For instance, 1, 2, and 11 are palindromic, and their cubes 1, 8, and 1331 are palindromic, too.

EXERCISE 127

1. Find five triangular numbers greater than 55 that are palindromic.
2. Find three palindromic numbers >11 whose squares are palindromic numbers.
3. Find three palindromic numbers >11 whose cubes are palindromic numbers.
4. Which of the following squares are palindromic numbers?

   a) 836^2 b) 1024^2 c) 10201^2
   d) 798644^2 e) 2112111^2 f) 64030648^2

Palindromic Primes and Emirps

Palindromic primes are numbers that are both prime and palindromic.

The first nine palindromic primes are 2, 3, 5, 7, 11, 101, 131, 151, and 181. It is not known if there are an unlimited number of palindromic primes.

> *Emirp* (prime spelled backwards) numbers are prime numbers that produce a different prime number when the digits are reversed.

EXAMPLE

The number 13 is a prime, and 31, the reverse of 13, is also a prime. So 13 and 31 are each called emirps.

EXAMPLE

The number 1033 is a prime, and 3301, the reverse of 1033, is also a prime. So 1033 and 3301 are emirps.

EXAMPLE

The cycles of 113, namely 113, 131, and 311 are primes. Then 113 and 311 are emirps, and 131 is a palindromic prime.

EXERCISE 128

1. Name five palindromic primes larger than 181.

2. a) Find four pairs of two-digit emirps.
   b) Find six pairs of three-digit emirps.
   c) Find six pairs of four-digit emirps.
   d) Find six pairs of five-digit emirps.

3. Find two three-digit numbers whose cycles are either emirps or palindromic primes.

4. **Challenge** Starting with the year 1000 A.D., name all emirp years through the end of the 20th century.

Honest Numbers

> An *honest* number in a language is a number whose word letter count and size are equal.

Honest numbers are evident in many of the approximately 6000 languages currently in use.

The reader is invited to research and uncover examples of honest numbers. For starters, here are some samples.

Language	Number	Number word
English	4	four
Spanish	5	cinco
Italian	3	tre

Try the two **Investigations: Honest Number Hunt** and **Seeking Honesty in Numbers.**

Bell Numbers

Eric Temple Bell (1883–1960) was a prolific and witty author who made a considerable contribution to our mathematical heritage through his stories that humanized mathematicians.

The numbers in the sequence $1, 2, 5, 15, \ldots$ are called Bell numbers in honor of this Scottish-born American mathematician because it was Bell's work that has brought significance to this set of numbers. Bell numbers can be found easily using the Bell triangle. In fact, the Bell numbers are the *last* number in each row of the Bell triangle that appears in the table on p. 199.

Note the last column of the table names the first four bell numbers $B_1 = 1$, $B_2 = 2$, $B_3 = 5$, $B_4 = 15$.

EXERCISE 129

1. Continue to form rows of the Bell triangle and find the values of the Bell numbers B_5 through B_{12}.
2. Is the number of rhyming patterns in a poem a Bell number? Study the chart below.

Number of lines in poem	Number of possible rhyming patterns
1	1
2	2: Lines 1 and 2 rhyme or they do not.
3	5: Lines 1, 2, and 3 rhyme
	Lines 1 and 2 (but not 3) rhyme.
	Lines 1 and 3 (but not 2) rhyme.
	Lines 2 and 3 (but not 1) rhyme.
	Lines 1, 2, and 3 do not rhyme in any way.

List the possible rhyme patterns for a four-line poem.

	Row	Bell triangle	Bell number
Step 1 Start with a 1 in row 1. Place 1 in row 2 under the 1 and add. Write the sum 2 beside 1 in row 2.	1	1	1
	2	1 2	2
Step 2 The last number in row 2 becomes the first in row 3. Add 2 and the number 1 above it. Write the sum 3 beside 2. Add 3 and the number 2 above it. Write the sum 5 beside 3.	3	2 3 5	5
Step 3 The last number in row 3 becomes the first in row 4. Add 5 and the number 2 above it. Write the sum 7 beside 5. Add 7 and the number 3 above it. Write the sum 10 beside 7. Write the sum 15 beside 10.	4	5 7 10 15	15

Bell Triangle

3. At the Penny Car Rental Co., the following information is on file. Study the chart to see if Bell numbers may be involved.

Number of customers	Number of ways customers can drive off in Penny cars
1	1
2	2: The two customers can ride separately in two cars or together in one car.
3	5: The three customers can ride together. Customers 1 and 2 (but not 3) can ride together. Customers 1 and 3 (but not 2) can ride together. Customers 2 and 3 (but not 1) can ride together. The three customers can ride separately.

List the possible ways four customers can ride in Penny cars.

4. You have three plates that are alike and 3 donuts: jelly, chocolate, and honey-dipped. In how many ways can you place the donuts on the plates? How many arrangements are there when you add a fourth kind of donut, say a coconut cream, and a fourth plate?

It should be noted that some descriptions and applications of the Bell numbers yield the sequence $B_0 = 1$, $B_1 = 1$, $B_2 = 2$, $B_3 = 5$, and so on.

Catalan Numbers

Before giving a definition of Catalan numbers, it is helpful to recall the meaning of some geometric terms.

A *convex polygon* is a polygon such that any two points in its polygonal region can be connected by a straight line segment that does not intersect the polygon.

Convex polygons

In a *nonconvex* or *concave polygon,* at least two points can be found in its polygonal region such that a straight line segment connecting the two points intersects two or more sides of the polygon.

Concave polygons

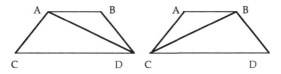

A *diagonal* of a convex polygon is any straight line segment that connects two nonadjacent vertices of the polygon.

Trapezoid *ABCD* has diagonals *AD* and *BC*

A *fixed* convex polygon is one that cannot be rotated, reflected, or translated to another position.

Leonhard Euler came upon the numbers that later were to be known as Catalan numbers while determining a solution to the problem "In how many ways can a fixed convex polygon be divided into triangles by drawing diagonals that do not intersect?"

EXAMPLE

The convex polygon with the fewest sides is a triangle. A triangle has no diagonals. The number of ways of dividing any triangle into triangles by nonintersecting diagonals is said to be 1. Thus, $C_1 = 1$.

EXAMPLE

A square or any convex quadrilateral has two diagonals.

There are 2 ways of dividing such figures into triangles with nonintersecting diagonals. So $C_2 = 2$.

EXAMPLE

A regular pentagon or any convex pentagon has five diagonals. There are five ways of dividing convex pentagons into triangles by nonintersecting diagonals.

The first three Catalan numbers are $C_1 = 1, C_2 = 2, C_3 = 5$.

Euler corresponded about this problem with his good friend and mentor, Christian Goldbach. He also involved his contemporary, the Hungarian, Johann Andreas von Segner (1704–1777) who spent most of his professional life in Germany. Von Segner discovered and published a routine for finding new Catalan numbers in terms of those already known. This was acknowledged as the earliest printed work that treated these numbers.

<u>von Segner's algorithm for finding Catalan numbers</u>

EXAMPLE

1. Start with 1. Follow it with $C_1 = 1$. That is, write

    1   1.

2. Reverse the numbers and write them under 1   1.

    1   1

    1   1

3. Multiply the numbers in pairs vertically.

$$\begin{array}{rr} 1 & 1 \\ \times\ 1 & \times\ 1 \\ \hline 1\ + & 1 = 2 = C_2 \end{array}$$

Then add the products together to obtain the second Catalan number.

EXAMPLE

1. Start with 1, followed by $C_1 = 1$ and $C_2 = 2$. That is, write

$$1 \qquad 1 \qquad 2.$$

2. Reverse the numbers and write them under 1 1 2.

$$\begin{array}{ccc} 1 & 1 & 2 \\ 2 & 1 & 1 \end{array}$$

3. Multiply the numbers in pairs vertically. Then add the products together to obtain the third Catalan number.

$$\begin{array}{rrr} 1 & 1 & 2 \\ \times\ 2 & \times\ 1 & \times\ 1 \\ \hline 2\ + & 1\ + & 2 = 5 = C_3 \end{array}$$

The Catalan numbers are named after Eugene Charles Catalan (1814–1894), who was born in Bruges, Belgium at a time when it was under French rule. He was professionally active in France and in Belgium. In 1838, Catalan solved a problem that counted the number of ways parentheses can be placed around a string of letters.

Try the **Investigation: Catalan Capers.**

It should also be noted that some descriptions and applications of the Catalan numbers yield $C_0 = 1$, $C_1 = 1$, $C_2 = 2$, $C_3 = 5$, and so on.

EXERCISE 130

1. Draw all the ways that a convex hexagon can be divided into triangles by drawing nonintersecting diagonals. Give the value of C_4, the total number of ways this can be done.
2. Use von Segner's method to compute the Catalan numbers from C_4 through C_{11}.
3. An even number of people are seated at a round table. Each person at the table extends one hand in a handshake and all people at the table clasp hands in pairs. No pair of joined hands may cross over any other pair of joined hands. Given the number of people, can you find in how many ways the handshakes can take place? Study the illustrations and then draw the number of ways of performing handshakes for six people and for eight people.

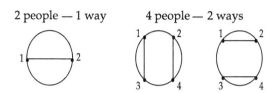

4. Six dancers, all of different heights stand in two rows for a picture. Each dancer in the front row must be shorter than his or her counterpart in the second row. In each row the dancers are arranged from shortest to tallest. How many ways can the dancers stand? What happens if there are 8 dancers? Sketch the arrangements for each case.

Try the **Investigations: Designing Designs, Geoboard Journeys** and **Mysterious Mountains & Binary Trees.**

Fibonacci Numbers

Leonardo of Pisa (c. 1170–1250) or Fibonacci as he was called was born in Italy and educated in Algeria. On his return to Italy, he wrote the **Liber Abaci** (1202), a book that introduced the Hindu–Arabic place-value decimal system and the use of Hindu–Arabic numerals to Europe. He was considered by many as the foremost mathematician of the Middle Ages.

Fibonacci wrote in **Liber Abaci**:

> *"When my father, who had been appointed by his country as public notary in the customs at Bugia acting for the Pisan merchants going there, was in charge, he summoned me to him while I was still a child, and having an eye to usefulness and future convenience, desired me to stay there and receive instruction in the school of accounting. There, when I had been introduced to the art of the Indians' nine symbols through remarkable teaching, knowledge of the art very soon pleased me above all else and I came to understand it, . . . "*

Liber Abaci contained a problem with a solution that paved the way for numerous discoveries and more problems and research that continues into the present time.

Here is that famous problem:

Given one adult pair of rabbits, determine how many pairs of rabbits there will be after one year if it is assumed

1. Every month each adult pair produces one new pair of babies.
2. The baby pairs grow to adulthood in one month. Then two months after their birth, each pair produces one pair per month.
3. No pairs are lost.

The pairs count is illustrated in the following tree diagram.

The Fibonacci Tree

The pairs count from the diagram is given in the table below.

Month	Adult pairs	Baby pairs	Total pairs
January	1	0	1
February	1	1	2
March	2	1	3
April	3	2	5
May	5	3	8
June	8	5	13

The second column in the table, Adult Pairs, yields the sequence of numbers $1, 1, 2, 3, 5, \ldots$.

These numbers are now called Fibonacci numbers. Let F_1 represent the first Fibonacci number, F_2 the second Fibonacci number, and so on.

The terms in the *Fibonacci sequence* are									
F_1,	F_2,	F_3,	F_4,	F_5,	F_6,	F_7,	F_8,	F_9,	$F_{10}, \ldots$
1,	1,	2,	3,	5,	8,	13,	21,	34,	$55, \ldots$

Here each term, excepting the first two, is the sum of the preceding two terms.

$$2 = 1 + 1,$$
$$3 = 1 + 2,$$
$$5 = 2 + 3, \quad \text{and so on.}$$

The *Fibonacci sequence* is the sequence F_n, with $F_1 = F_2 = 1$, and $F_n = F_{n-1} + F_{n-2}$ for $n > 2$.

> A *Fibonacci-like sequence* is a sequence that has a recursive rule of formation like the Fibonacci sequence. Each term, excepting the first two terms, is the sum of the preceding two terms.

EXAMPLE

Create a Fibonacci-like sequence with the first two terms 1, 4.

1, 4, 5, 9, 14, 23, 37, 60, 97, 157, ... is a Fibonacci-like sequence because $5 = 1 + 4, 9 = 4 + 5, 14 = 5 + 9, 23 = 9 + 14$, and so on.

Fibonacci and Fibonacci-like sequences share an interesting property. List the first ten terms of the sequences. Study the following illustration.

Term	Fibonacci numbers	Fibonacci-like numbers
1	1	1
2	1	4
3	2	5
4	3	9
5	5	14
6	8	23
7	13	37
8	21	60
9	34	97
10	$\underline{55}$	$\underline{157}$
	$143 = 11(13)$	$407 = 11(37)$

The sum of the first ten terms of both the Fibonacci and the Fibonacci-like sequence equals 11 times the 7th term in that sequence.

EXERCISE 131

1. Starting with F_{11}, name the next ten Fibonacci numbers.
2. Do you think that the number of Fibonacci numbers is unlimited? Give a reason for your answer.
3. Verify that the sum of the first ten numbers in the Fibonacci-like sequence starting with 5, 6 equals 11 times the 7th term in the sequence.
4. Show that the sum of the first ten terms in *any* Fibonacci-like sequence is 11 times the 7th term. Explain your answer.

5. Study the patterns in the table. Then extend the table through the sum of the first ten Fibonacci numbers.

The sum of the first	Fibonacci numbers equals	It is found by subtracting	From the Fibonacci number
2	2	1	$F_4 = 3$
3	4	1	$F_5 = 5$
4	7	1	$F_6 = 8$

6. Describe a way to find the sum of the first n Fibonacci numbers.

Try the **Investigation: Fibonacci Fascinations.**

Lucas Numbers

One of the more well-known Fibonacci-like sequences of numbers is named after Francois Edouard Anatole Lucas (1842–1891), a French mathematician. Lucas studied the sequence 1, 1, 2, 3, 5, 8, ... and in fact, he was the one who named it the *Fibonacci* sequence.

The terms in the *Lucas sequence* are
L_1, L_2, L_3, L_4, L_5, L_6, L_7, L_8, L_9, L_{10}, ...
1, 3, 4, 7, 11, 18, 29, 47, 76, 123, ...

EXERCISE 132

1. Study the patterns in the table below. Then extend the table through the sum of the first nine Lucas numbers.

The sum of the first	Lucas numbers equals	It is found by subtracting	From the Lucas number
2	4	3	$L_4 = 7$
3	8	3	$L_5 = 11$
4	15	3	$L_6 = 18$

2. Describe a way to find the sum of the first n Lucas numbers.

Tribonacci Numbers

Tribonacci numbers are the result of another extension of the addition pattern in the Fibonacci sequence.

The terms in the *tribonacci sequence* are
T_1, T_2, T_3, T_4, T_5, T_6, T_7, T_8, T_9, T_{10}, ...   1, 1, 2, 4, 7, 13, 24, 44, 81, 149, ...

Here each term, excepting the first three, is the sum of the preceding three terms.

$4 = 1 + 1 + 2$ $7 = 1 + 2 + 4,$
$13 = 2 + 4 + 7$ $24 = 4 + 7 + 13,$ and so on.

> The *tribonacci sequence* is the sequence $1, 1, 2, 4, 7, ...$ in which each term, excepting the first three terms, is the sum of the preceding three terms.
>
> A *tribonacci-like sequence* is a sequence that has an additive rule of formation like the tribonacci sequence. Each term, excepting the first three terms, is the sum of the preceding three terms.

EXAMPLE

The set of numbers 1, 2, 3, 6, 11, 20, 37, 68, 125, ... is a tribonacci-like sequence because each term except the first three is the sum of the preceding three terms.

An interesting application of the tribonacci numbers appears in the solution of the following problem.

EXAMPLE

The Wondersquare Problem

Take four nonnegative integers, form cyclic differences of pairs of numbers (smaller from larger). Make the fourth difference using the last number and the first number. Repeat the process as needed to achieve a goal of all zeros.

Is it possible to give four starting numbers that yield all zeros at any desired stage?

Form the cyclic differences of pairs of numbers in the set 9, 4, 6, 7. Repeat the process until all zeros are reached.

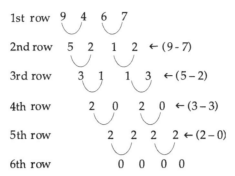

1st row 9 4 6 7

2nd row 5 2 1 2 ← (9 - 7)

3rd row 3 1 1 3 ← (5 - 2)

4th row 2 0 2 0 ← (3 - 3)

5th row 2 2 2 2 ← (2 - 0)

6th row 0 0 0 0

Starting with 9, 4, 6, 7, and following the procedure described, the process ends at the 6th row with all zeros.

EXERCISE 133

1. Starting with T_{11}, name the next ten tribonacci numbers.
2. Investigate the Wondersquare Problem. Form the cyclic differences of pairs of numbers in the sets below. Find the number of the row in which all differences are zero.

    a) 2, 0, 2, 4   b) 0, 2, 2, 4   c) 1, 1, 3, 5
    d) 0, 1, 2, 3   e) 1, 1, 2, 4   f) 0, 1, 2, 4
    g) 0, 7, 20, 44   h) 2, 3, 6, 11   i) 24, 44, 81,149
    j) 2, 4, 7, 13   k) 0, 2, 6, 13   l) 0, 24, 68, 149
    m) 6, 11, 20, 37   n) 7, 13, 24, 44   o) 20, 37, 68, 125

3. Consider several sequences of four consecutive even numbers. Find the number of the row in which all differences are zero. Is there a pattern in your answers?
4. Consider several sequences of four consecutive multiples of 5. Find the number of the row in which all differences are zero. Is there a pattern in your answers?
5. **Challenge** Try the Wondersquare Problem with other special sequences, look for a connection with the tribonacci numbers, and write a report on your findings.

Tetranacci Numbers

These numbers are the result of yet another extension of the addition pattern in the Fibonacci and tribonacci sequences.

The terms in the *tetranacci sequence* are										
Q_1,	Q_2,	Q_3,	Q_4,	Q_5,	Q_6,	Q_7,	Q_8,	Q_9,	Q_{10},	...
1,	1,	2,	4,	8,	15,	29,	56,	108,	208,	...

Here each term, excepting the first four, is the sum of the preceding four terms.

$$8 = 1 + 1 + 2 + 4$$

$$15 = 1 + 2 + 4 + 8, \text{ and so on.}$$

> The *tetranacci sequence* is the sequence 1, 1, 2, 4, 8, 15, . . . in which each term, excepting the first four terms, is the sum of the preceding four terms.
>
> A *tetranacci-like sequence* is a sequence that has an additive rule of formation like the tetranacci sequence. Each term, excepting the first four terms, is the sum of the preceding four terms.

Phibonacci Numbers

In 1980, Anders Bager of Denmark made the following observation about Euler ϕ numbers described on p. 47. In some instances the Euler ϕ number equals the sum of the preceding two Euler ϕ numbers. He called such numbers *phibonacci* numbers.

EXAMPLE

Is 3 a phibonacci number?

That is, does $\phi(3) = \phi(2) + \phi(1)$?

$\phi(3) = 2$, $\phi(2) = 1$, $\phi(1) = 1$, and $2 = 1 + 1$.

So 3 is a phibonacci number.

EXAMPLE

Is 5 a phibonacci number?

That is, does $\phi(5) = \phi(4) + \phi(3)$?

$\phi(5) = 4$, $\phi(4) = 2$, $\phi(3) = 2$ and $4 = 2 + 2$.

So 5 is a phibonacci number.

EXERCISE 134

1. Starting with Q_{11} name the next ten tetranacci numbers.
2. Which of the following integers are phibonacci numbers?

   a) 6   b) 7   c) 9   d) 11   e) 17   f) 23   g) 29

3. Given that 1 and 2 are the first two terms in the phibonacci number sequence, name the next eight terms in the sequence.

Survivor Numbers or U-Numbers or Ulam Numbers

> A number x is called a *survivor* number if x can be written as the sum of two different smaller survivor numbers in exactly one way. The numbers 1 and 2 are given as survivor numbers.

Survivor numbers are well known under the name Ulam numbers or U-numbers. Stanislaw Ulam introduced them in 1964. They have generated considerable interest and stimulated numerous questions and extensions.

EXAMPLE

Find the next four survivor numbers after 1 and 2.

$1 + 2 = 3$	Therefore, 3 is a survivor.
$3 + 1 = 4$	Therefore, 4 is a survivor.
$3 + 2 = 5$ and $4 + 1 = 5$	Therefore, 5 is *not* a survivor.
$4 + 2 = 6$	Therefore, 6 is a survivor.
$4 + 3 = 7$ and $6 + 1 = 7$	Therefore, 7 is *not* a survivor.
$6 + 2 = 8$	Therefore, 8 is a survivor.

Thus, the next four survivor numbers are 3, 4, 6 and 8.

EXAMPLE

Consider 9. Is it a survivor?

The first six survivors are 1, 2, 3, 4, 6, and 8. Since 9 can be written as the sum of these survivor numbers in more than one way, $1 + 8 = 9$ and $3 + 6 = 9$, 9 is not a survivor.

EXERCISE 135

1. Which of the following numbers are survivors?

   a) 26   b) 27   c) 28

2. Find some examples amongst pairs of two-digit survivor numbers that show the

   a) sum of two survivor numbers may be a survivor number.
   b) difference of two survivor numbers may be a survivor number.
   c) product of two survivor numbers may be a survivor number.
   d) quotient of two survivor numbers may be a survivor number.

Tautonymic Numbers

In each of the following words, two or more letters are repeated.

MAMA   PAPA   MURMUR   BONBON   HULAHULA

Words consisting of repeating groups of letters are called *word tautonyms*. Similarly, there are *tautonymic numbers* in which the number's digits are repeated.

EXAMPLE

The numbers 22 and 444 are tautonymic numbers with one digit that repeats.

EXAMPLE

The numbers 2424 and 363636 are tautonymic numbers with two digits that repeat.

Tautonymic numbers are involved in interesting computation patterns, some of which are shown below.

EXAMPLE

$$101 \times 59 = 5959$$
$$101 \times 13 = 1313$$
$$101 \times 29 = 2929$$

$$1001 \times 121 = 121121$$
$$1001 \times 235 = 235235$$

EXAMPLE

$$\underline{175}175 \div 7 = 25025$$
$$25025 \div 11 = 2275$$
$$2275 \div 13 = \underline{175}$$

EXERCISE 136

1. Find each of the following products.

   a) 101×38 b) 101×46 c) 101×99
   d) 1001×134 e) 1001×240 f) 1001×999
   g) $251 \times 7 \times 11 \times 13$ h) $365 \times 7 \times 11 \times 13$

2. If a, b, c represent any digits, what is $abc \times 7 \times 11 \times 13$?
3. Divide each of the following numbers successively by 7, by 11, and by 13 as shown in the example above.

   a) 237237   b) 456456   c) 689689   d) 888888   e) 946946

4. If a, b, c represent any selection of three digits except 0, 0, 0, does $(((abcabc \div 7) \div 11) \div 13) = abc$? Explain your answer.

Lagado Numbers

According to the prime factorization property, each composite number can be factored into a product of primes in exactly one way. The sequence of numbers 1, 4, 7, 10, 13, 16, 19, 22, 25, 28, 31, 34, 37, 40, 43, 46, 49, 52, 55, 58, 61, 64, ... are called *Lagado numbers*. Lagado numbers appear in Jonathan Swift's **Gulliver's Travels**.

A *Lagado prime* is a Lagado number greater than 1 that is not the product of any Lagado number except 1 and itself.

The first five Lagado primes are 4, 7, 10, 13, and 19.

EXAMPLE

Is 28 a Lagado prime?

$28 = 4 \times 7$, where 4 and 7 are Lagado numbers.

So 28 is not a Lagado prime.

What about prime factorization in the set of Lagado numbers?

EXAMPLE

Factor the Lagado composite number 40 into Lagado primes.

$$40 = 4 \times 10$$

40 can be factored into Lagado primes in exactly one way.

EXAMPLE

Factor the Lagado composite number 100 into Lagado primes.

$$100 = 4 \times 25 \qquad 100 = 10 \times 10$$

100 can be factored into Lagado primes in two distinct ways.

Thus, the Lagado number system does *not* have a unique factorization property for its composite numbers.

EXERCISE 137

1. Factor the following Lagado composite numbers in as many ways as possible.

   a) 130   b) 154   c) 220   d) 232
   e) 376   f) 400   g) 460   h) 484

2. Give a rule that generates the Lagado numbers.

   The Lagado numbers are the last family to be introduced in this collection.

*****

The reader is challenged to continue working with number families on three levels:

1. Seek additional information about the number families you have met on these pages. What other properties do they have and what problems have been or are yet to be solved about them?

2. Search for different number families to add to your collection — provide some facts about their origin and give some details about their properties and related problems.

3. Create your own number family — describe its members, identify their characteristics, properties, and applications!

Echoing the words of the French mathematician, scientist, and philosopher —
" ... (we) hope that posterity will judge (us) kindly not only as to the things which (we) have explained but also as to those which (we) have omitted so as to leave to (you) the pleasure of discovery."

René Descartes (1596–1650)

Recommended Readings

Beiler, Albert. **Recreations in the Theory of Numbers — The Queen of Mathematics Entertains**. New York, NY: Dover Publications, Inc., 1964, (paperback, 1999).

Conway, John & Guy, Richard. **The Book of Numbers**. New York, NY: Springer-Verlag, 1996.

Freidberg, Richard. **An Adventurer's Guide to Number Theory**. New York, NY: Dover Publications, Inc., 1994, (paperback, 2003).

Gardner, Martin. **Martin Gardners Mathematical Games: The Entire Collection of His** *Scientific American* **Columns on One CD**. Washington, D.C.: The Mathematical Association of America, 2005.

Guy, Richard K. **Unsolved Problems in Number Theory,** 3rd edition. New York, NY: Springer Verlag, 2004.

Joseph, George G. **The Crest of the Peacock: Non-European Roots of Mathematics,** 3rd edition. Princeton, NJ: Princeton University Press, 2010.

Kanigel, Robert. **The Man Who Knew Infinity: A Life of the Genius**, **Ramanujan.** New York, NY: Washington Square Press, 1992.

Koshy, Thomas. **Triangular Arrays with Applications**. New York, NY: Oxford University Press, 2011.

Posamentier, Alfred S. and Lehmann, Ingmar. **The Fabulous Fibonacci Numbers.** Amherst, NY: Prometheus Books, 2007.

Swetz, Frank. **Legacy of the Luoshu: The 4,000 Year Search for the Meaning of the Magic Square of Order Three**. Natick, MA: A.K. Peters/CRC Press, 2008.

Wells, David. **The Penguin Dictionary of Curious & Interesting Numbers.** New York, NY: Penguin Group USA, 1986. (paperback, 1998).

Useful Online Resources

MacTutor History of Mathematics. www-history.mcs.st-and.ac.uk
The Online Encyclopedia of Integer Sequences. (OEIS) www.oeis.org
The Prime Pages. www.primes.utm.edu
Wikipedia, the Free Encyclopedia. www.wikipedia.org
Wolfram Math World. www.mathworld.wolfram.com

Glossary of Numbers

Why are numbers beautiful? It's like asking why is Beethoven's Ninth Symphony beautiful. If you don't see why, someone can't tell you. I know numbers are beautiful. If they aren't beautiful, nothing is.

Paul Erdos (1913–1996)

abundant (excessive, redundant) A number for which the sum of its proper divisors is greater than the number.

additive digital root An integer 1 through 9 found by summing the digits of the given number, repeatedly if necessary.

additive multidigital A number divisible by the sum of its digits.

additive persistence The number of steps required to find the additive digital root of a given integer when adding digit by digit.

aliquot parts See proper divisors.

almost perfect A number with sum of its proper divisors 1 more or less than the number.

amicable digital invariant pair A pair of numbers for which the number chain of each number ends in the other.

amicable pair (friendly, sympathetic) A pair of numbers for which each number is the sum of the proper divisors of the other.

Armstrong (perfect digital invariant) An n-digit number that equals the sum of the nth powers of its digits.

balanced A number n for which there are as many composite numbers between 1 and n as there are prime numbers.

baselike set A set such that all numbers less than or equal to a given number can be written as a sum of numbers in the set.

Bell The numbers in the sequence (1), 1, 2, 5, 15, 52, 203, 877, 4140, ...

beprisque A number adjacent to a prime and a square or vice versa.

bigrade pair of triples A pair of triples for which the sum of the numbers of one triple equals the sum of the numbers of the other triple and the sum of the squares of the numbers of one triple equals the sum of the squares of the numbers of the other triple.

binomial coefficient A number that can be expressed in the form $\dfrac{n!}{r!(n-r)!}$ where n is a nonnegative integer and $r = 0, 1, \ldots, n$.

Boolean A number whose prime factorization contains only distinct primes raised to the first power.

Catalan The numbers in the sequence $(1), 1, 2, 5, 14, 42, \ldots$, or numbers that can be expressed in the form $\dfrac{(2n)!}{n!(n+1)!}$ with $n = 0, 1, 2, \ldots$.

centered square The numbers in the sequence $1, 5, 13, 25, 41, \ldots$, or numbers that can be expressed in the form $2n^2 - 2n + 1$ with $n \geq 1$.

centered triangular The numbers in the sequence $1, 4, 10, 19, 31, \ldots$, or numbers that can be expressed in the form $\dfrac{(3n^2 - 3n + 2)}{2}$ with $n \geq 1$.

composite A number greater than 1 with more than two divisors.

congruent A number that is twice the product of the two smaller numbers in a Pythagorean triple.

consecutive set Numbers that follow one another in some sequence.

co-prime pair Numbers that are relatively prime, that is, have only the common divisor 1.

crowd A number whose chain of sums of proper divisors that return to the original number has 3 links.

cubic A number that can be expressed in the form n^3.

Cullen A number that can be expressed in the form $n(2^n) + 1, n \geq 0$.

defective See deficient.

deficient (defective) A number that is greater than the sum of its proper divisors.

divisor (exact divisor) An integer that divides a given integer with remainder zero.

doubling The specified doubling-subtraction procedure eventually ends in a 1 and the total number of doublings is 1 less than the starting number.

doubly perfect The sum of the divisors of the number equals twice the number.

emirp A prime number whose reversal is a different prime.

Euler phi The number of integers less than or equal to a given number that are relatively prime to that number.

even A number that can be expressed in the form $2n$.

even-type A number for which the number of primes in its factorization is even.

exact divisor See divisor.

excessive See abundant.

extremely modest A number which is modest with respect to every allowable sectioning of itself.

factor An exact divisor of a number.

factorial A number that can be expressed in the form $n! = 1 \times 2 \times 3 \times \cdots \times n$, with $0! = 1$.

factorial sum An integer that equals the sum of the factorials of its digits.

female An even number.

Fermat A number that can be expressed in the form $2^{2^n} + 1$, $n \geq 0$.

Fibonacci The numbers in the sequence $1, 1, 2, 3, 5, 8, \ldots$, in which each new term, excepting the first two terms, is the sum of the preceding two terms.

Fibonacci-like A sequence that has an additive rule of formation like the Fibonacci sequence. Each term, excepting the first two terms, is the sum of the two preceding terms.

figurate (polygonal) A number that can be represented by dots in specific geometrical or polygonal patterns.

friendly See amicable pair.

Germain prime A prime p for which $2p + 1$ is prime.

gnomic A term of an arithmetic progression; terms that differ from each other by a constant.

good A number $n = a_1 + a_2 + \cdots + a_k$ such that $\dfrac{1}{a_1} + \dfrac{1}{a_2} + \cdots + \dfrac{1}{a_k} = 1$ with $k < n$.

greatest common divisor The largest integer that divides two or more integers.

hailstone A number for which a result of 1 is reached after following the rules: a) if the number is odd, multiply it by 3 and add 1, and b) if the number is even, divide it by 2.

happy A number for which the sum of the squares of its digits, taken repeatedly if necessary, is 1.

heptagonal A number that can be represented by dots in the form of a regular heptagon, or expressed in the form $\frac{n(5n-3)}{2}$.

heptagonal pyramidal A number that can be written as a sum of the first k heptagonal numbers.

hexagonal A number that can be represented by dots in the form of a regular hexagon, or expressed in the form $n(2n-1)$.

hexagonal pyramidal A number that can be written as a sum of the first k hexagonal numbers.

highly composite An integer n greater than 1 that has more divisors than any integer less than n.

honest A number whose word letter count and size are equal.

imperfectly amicable pair A pair of numbers for which the sum of the proper divisors of each number is the same.

integer A number in the sequence $\ldots, -3, -2, -1, 0, 1, 2, 3, \ldots$. In this book integer means positive integer $1, 2, 3, \ldots$.

integral divisor See divisor.

Kaprekar's number The number 6174.

Lagado A number that can be expressed in the form $3n - 2$.

Langford sequence For the sequence (k, n), a rearrangement of k copies of the integers from 1 to n such that between each occurrence of the integer j there are exactly j integers.

least common multiple The smallest integer that can be divided by each number in a given pair or group.

linear A positive integer.

lonely A number that does not equal the sum of any number plus that number's digits.

Lucas The numbers in the sequence $1, 3, 4, 7, 11, 18, \ldots$, in which each new term, excepting the first two terms, is the sum of the preceding two terms.

lucky The numbers $1, 3, 7, 9, \ldots$, determined by a sieve process similar to the prime sieve of Eratosthenes.

male An odd number greater than 1.

Mersenne prime A prime of the form $2^n - 1$, where n is prime.

modest A number greater than 10 that can be sectioned into two parts, having the property that when the given number is divided by the second part, the remainder is the first part.

monodigit A number formed using only one distinct digit.

multiperfect (multiply perfect, pluperfect) A number for which the sum of its divisors equals a multiple of the number.

multiple An integer that results from multiplying a given integer by another integer.

multiplicative digital root One of the integers $0, 1, 2, \ldots, 9$ which corresponds to the result of multiplying the digits of a given integer, repeatedly if necessary.

multiplicative multidigital A number that is divisible by the product of its digits.

multiplicative persistence The number of steps required to find the multiplicative digital root number of a given integer.

multiply perfect See multiperfect.

narcissistic A number that can be represented by taking its digits in order in combination with some mathematical operation(s).

nearly good semiperfect A semiperfect number n such that $n = a_1 + a_2 + \cdots + a_k$ with a_i divisors of n and such that $\dfrac{a_1}{n} + \dfrac{a_2}{n} + \cdots + \dfrac{a_k}{n} = 1$ with $k < n$.

nude A number that is divisible by each of its nonzero digits. If the number has more than two digits then it must be divisible by at least one other number, other than the number itself, formed of two or more digits of the number.

oblong See rectangular.

octagonal (star, stellate) A number that can be represented by dots in the form of a regular octagon, or expressed in the form $n(3n - 2)$.

octagonal pyramidal A number that can be written as a sum of the first k octagonal numbers.

odd A number that can be expressed in the form $2n - 1$.

odd-type A number for which the number of primes in its factorization is odd.

palindrome A number that reads the same backward or forward.

palindromic prime A number that is both prime and palindromic.

pentagonal A number that can be represented by dots in the form of a regular pentagon, or ex-pressed in the form $n(3n - 1)/2$.

pentagonal pyramidal A number that can be written as a sum of the first k pentagonal numbers.

perfect A number that equals the sum of its proper divisors.

perfect digital invariant See Armstrong.

phibonacci An Euler phi number that is equal to the sum of the preceding two Euler phi numbers.

plane See rectangular.

pluperfect See multiperfect.

polygonal See figurate.

positive integers The numbers 1, 2, 3, ...

positive prime pair A number n such that it is possible to pair off the numbers from 1 to n with the numbers 1 to n so that the sum of each pair is a prime.

positive square pair A number n such that it is possible to pair off the numbers from 1 to n with the numbers 1 to n so that the sum of each pair is a square.

powerful A number that can be written as a sum of positive integer powers of its digits.

practical A number for which each number less than or equal to it is a divisor of the number or can be written as a sum of distinct divisors of the number.

prime A number with exactly two divisors.

prime circle A number n such that the numbers from 1 to n can be arranged in a circle so that the sum of any two adjacent numbers is prime.

prime factor A factor of an integer that is a prime number.

prime line A number n such that the numbers from 1 to n can be arranged in a row so that the sum of any two adjacent numbers is prime.

primitive Pythagorean triple A Pythagorean triple in which the numbers have only the common factor 1.

primitive semiperfect A semiperfect number that is not divisible by any other semiperfect number.

proper divisors (aliquot parts) Divisors of an integer, not including the integer itself.

pyramidal A number that can be written as a sum of the first k figurate numbers of a given type.

Pythagorean parallelepiped (quadruple) The sum of the squares of the first three numbers in the quadruple equals the square of the fourth number.

Pythagorean quadruple See Pythagorean parallelepiped.

Pythagorean triple A triple in which the sum of the squares of the first two numbers equals the square of the third number.

rectangular (plane, oblong) A number that can be expressed in the form $n(n+m)$ with $n = 1, 2, 3, \ldots$ and $m = 2, 3, 4, \ldots$.

rectangular pyramidal A number that can be written as a sum of the first k rectangular numbers.

recurring digital invariant of the k th order A number such that the chain of numbers produced by taking the sums of the kth power of the digits returns to the number.

redundant See abundant.

relatively prime Two or more numbers that have the greatest common divisor of 1.

repunit A number formed using only 1s as digits.

Ruth-Aaron Two consecutive numbers n, $n+1$ for which $S(n) = S(n+1)$, where $S(n)$ is the sum of the product of each prime factor of n and its corresponding exponent.

semiperfect A number that equals the sum of some but not all of its proper divisors.

semiprime A number that is a product of two distinct primes.

series An indicated sum of a sequence of numbers.

snowball prime set The prime numbers formed by adding on additional digits one by one to the right of a given prime or 1.

sociable A number that has three or more links in the chain of sums of proper divisors that return to the given number.

social A number equal to the sum of a number plus that number's digits.

solid (space, spatial) The three-dimensional counterpart of some figurate number.

space See solid.

spatial See solid.

square A number that can be expressed in the form n^2.

squarefree A number that is not divisible by the square of any prime number.

square pyramidal A number that can be written as a sum of the first k square numbers.

star See octagonal.

star pyramidal A number that can be written as a sum of the first k star numbers.

stellate See octagonal.

subfactorial A number that can be expressed in the form

$$!n = n!\left(1 - \left(\frac{1}{1!}\right) + \left(\frac{1}{2!}\right) - \left(\frac{1}{3!}\right) + \cdots + (-1)^n \left(\frac{1}{n!}\right)\right).$$

survivor A number that can be written as the sum of two different smaller survivor numbers in exactly one way. 1 and 2 are the first two survivor numbers.

sympathetic See amicable pair.

tautonymic A number whose digits are repeated in order.

tetragonal A number that can be represented by dots in the form of a particular quadrilateral, or expressed in the form

$$\frac{(3n^2 - 3n + 2)}{2}.$$

tetrahedral A number that is the sum of the first k triangular numbers.

tetranacci The numbers in the sequence 1, 1, 2, 4, 8, 15, ..., in which each new term, excepting the first four terms, is the sum of the preceding four terms.

tetranacci-like A sequence that has an additive rule of formation like that of the tetranacci sequence. Each term, excepting the first four terms, is the sum of the preceding four terms.

trapezoidal A figurate number that can be represented by a trapezoidal array of two or more rows of dots or chips.

triangle inequality triple Three numbers that represent the lengths of the sides of a triangle.

triangular A number that can be represented in the form $\frac{n(n+1)}{2}$.

tribonacci The numbers in the sequence 1, 1, 2, 4, 7, 13, ..., in which each new term, excepting the first three terms, is the sum of the preceding three terms.

tribonacci-like A sequence that has an additive rule of formation like the tribonacci sequence. Each term, excepting the first three terms, is the sum of the preceding three terms.

triperfect A number for which the sum of its divisors equals three times the number.

twin primes Two primes with difference 2.

Ulam A number for which a quotient of 1 is reached after following the rules: a) if the number is odd, multiply it by 3, add 1, and divide the sum by 2, and b) if the number is even, divide it by 2.

visible factor A number that is divisible by each of its nonzero digits.

weird abundant An abundant number that is not semiperfect.

Solutions to Investigations

Catalan Capers

1. $C_3 = 5$

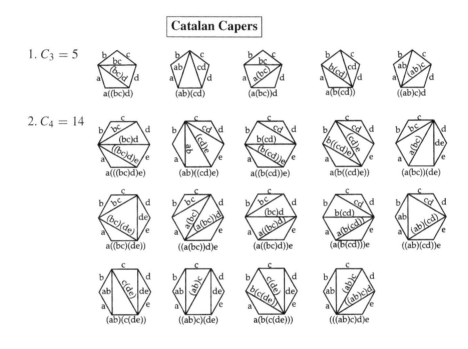

2. $C_4 = 14$

Centered Triangular Numbers

1. The figure for $s_6 = 46$ is shown on p. 225.

2.

n	1	2	3	4	5	6	7	8	9	10
Dot Count s_n	1	4	10	19	31	46	64	85	109	136

3. Recursive formula:
$s_1 = 1$,
$s_2 = 4 = s_1 + 3 \cdot 1$,
$s_1 = 1, s_n = s_{n-1} + 3(n-1)$,
$s_3 = 10 = s_2 + 3 \cdot 2$,
$s_4 = 19 = s_3 + 3 \cdot 3$
$n = 2, 3, 4, \ldots$

4. Explicit formula:
Recall the sequence of triangular numbers
$1, 3, 6, 10, \ldots$, where $t_n = \dfrac{n(n+1)}{2}$.
$s_1 = 1 = 1^2, s_2 = 4 = 2^2, s_3 = 10 = 3^2 + 1 = 3^2 + t_1$
$s_4 = 19 = 4^2 + 3 = 4^2 + t_2, s_5 = 31 = 5^2 + 6 = 5^2 + t_3$
$s_n = n^2 + t_{n-2} = n^2 + \dfrac{(n-1)(n-2)}{2}$
$= \dfrac{3n^2 - 3n + 2}{2}, n = 1, 2, 3, \ldots$

5.

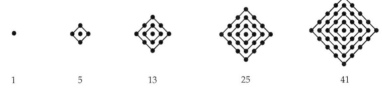

| 1 | 5 | 13 | 25 | 41 |

6.

n	1	2	3	4	5	6	7	8	9	10
Dot Count q_n	1	5	13	25	41	61	85	113	145	181

7. Recursive formula: $q_1 = 1, q_n = q_{n-1} + 4(n-1), n = 2, 3, 4, \ldots$.
8. Explicit formula: $q_n = 2n^2 - 2n + 1, n = 1, 2, 3, \ldots$.
   Also q_n can be written as $q_n = n^2 + (n-1)^2$

Conjecturing with Pascal

1. In row n, the row sum is 2^n. The cumulative row sum for rows 0 through n is $2^{n+1} - 1$.

2. For case $+/-$: In row n, the sum is 2 for $n > 0$. For case $-/+$: In row n, the sum is 0 for $n > 0$.
3. a,c) All numbers are odd in rows 0, 1, 3, 7, 15, 31, 63, 127, ... , $2^n - 1, n \geq 0$.
   b,c) All numbers are even, except for the two 1s in rows 2, 4, 8, 16, 32, 64, 128, ... , $2^n, n \geq 1$.
4.

Triangular numbers 1, 6, 28, 120

5. There are numerous helpful references. Consult NCTM Illuminations and Publications. Also check online — Fractal Foundation and others.

Crisscross Cubes

1.

2. Perimeter: $P_1 = 4$, $P_2 = 12$, $P_3 = 20$, $P_4 = 28$, $P_5 = 36$, $P_6 = 44$
   Volume: $V_1 = 1$, $V_2 = 5$, $V_3 = 13$, $V_4 = 25$, $V_5 = 41$, $V_6 = 61$
   TSA: $S_1 = 6$, $S_2 = 22$, $S_3 = 46$, $S_4 = 78$, $S_5 = 118$, $S_6 = 166$
3.

	Recursive formula	Explicit formula
Perimeter:	$P_1 = 4$, $P_n = P_{n-1} + 8, n > 1$	$P_n = 4(2n - 1)$
Volume:	$V_1 = 1$, $V_n = V_{n-1} + 4(n - 1), n > 1$	$V_n = 2n^2 - 2n + 1$
		$= n^2 + (n - 1)^2$
Total Surface Area:	$S_1 = 6$, $S_n = S_{n-1} + 8n, n > 1$	$S_n = 4n^2 + 4n - 2$

Dealing with Digits Base 10

1. There are 271 numbers between 0 and 1000 that have at least one 6. In all intervals of 0–99, 100–199, and so on, there are 19 numbers which contain at least one 6, except the interval 600–699 in which all 100 numbers have at least one 6.

2.

	4s	6s	9s
0–99	20	20	20
100–199	20	20	20
200–299	20	20	20
300–399	20	20	20
400–499	120	20	20
500–599	20	20	20
600–699	20	120	20
700–799	20	20	20
800–899	20	20	20
900–964	17	11	71
Total:	297	291	251

3.

Pages	Number of Pages	Number of Digits	Cumulative Count (Digits)
1–9	9	9	9
10–99	90	180	189
100–199	100	300	489
200–299	100	300	789
300–388	89	267	1056
Total:	388		

Complete book has 393 pages.

4. 121 numbers. There are 9 of the form "*aaa*", and 0 of the form "*aab*". There are 96 of the form '*abc*' (no zero). There are 16 different threesomes, and each threesome makes 6 numbers. Digits are 1 apart, such as 123; 2 apart, such as 246; or 3 apart, such as 369. There are also 16 of the form '*abc*' with one digit a zero. There are 4 threesomes: 0, 1, 2; 0, 2, 4; 0, 3, 6; and 0, 4, 8, and each threesome makes four numbers.

5. The 413th is 431,625. Since there are 5! = 120 numbers starting with a 1, 120 numbers starting with a 2 and another 120 numbers starting with a 3, giving a total of 360 numbers, the 413th number must begin with a 4. There are 24 numbers each beginning with 41 and 42, bringing the total to 408 numbers less than 43x,xxx. The next five numbers in order of size are: 431,256, 431,265, 431,526, 431,562, and 431,625.

6. See Conway, J. and Guy, R., **The Book of Numbers.** NY: Springer-Verlag, 1996.

7. See Joseph, G.G., **The Crest of the Peacock: Non-European Roots of Mathematics**, 3rd edition. NJ: Princeton University Press, 2010.

Designing Designs

1. Let 1, 2, 3, 4 be the four colors, then the 15 possible arrangements are:

One Color	Two Colors			
1111	1112	1211	1122	1212
	1121	2111	1221	

Three Colors			Four Colors
1231	2311	2113	1234
1123	2131	1213	

2. Let 1, 2, 3, 4, 5 be the five colors, then there are 52 possible arrangements and 54 chapters in the book.

One Color	Two Colors		
11111	11112	11122	21121
	11121	11221	21211
	11211	12211	12121
	12111	22111	12112
	21111	21112	11212

Three Colors				
11123	12311	11223	12123	21123
11213	21113	11232	12132	21132
11231	21131	11322	12312	21312
12113	21311	13122	13212	23112
12131	23111	31122	31212	32112

Four Colors					Five Colors
11234	12314	21134	21341	23141	12345
12134	12341	21314	23114	23411	

Factor Lattices

1.

a) $24 = 2^3 \cdot 3$ b) $72 = 2^3 \cdot 3^2$ c) $100 = 2^2 \cdot 5^2$

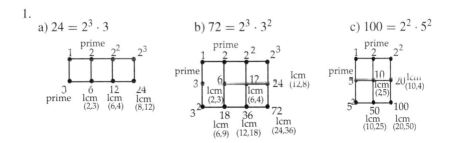

2. Answers will vary. Examples: $42 = 2 \cdot 3 \cdot 7$, $66 = 2 \cdot 3 \cdot 11$, $70 = 2 \cdot 5 \cdot 7$, $105 = 3 \cdot 5 \cdot 7$, $110 = 2 \cdot 5 \cdot 11$

3. Answers will vary. The number must be the product of 3 primes, one of which is squared, and the other two are raised to the first power.

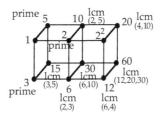

For a, b, c prime, $N = a^2 \cdot b \cdot c$.
For example, $60 = 2^3 \cdot 3 \cdot 5$

4. $90 = 2^3 \cdot 3^2 \cdot 5$ $168 = 2^3 \cdot 3^2 \cdot 7$

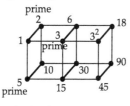   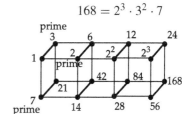

5. $210 = 2 \cdot 3 \cdot 5 \cdot 7$

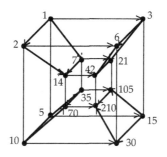

Factorial Finishes

1. Note that zeros are produced by 10s, which in turn are the product of 2s and 5s. Since there are many 2s in the factorials, find the number of 10s by counting the number of 5s. Consider

   $6! = 6 \cdot \underline{5} \cdot 4 \cdot 3 \cdot 2 \cdot 1 = 720$ (One five, so one zero.)

   a) 50! has 12 zeros. There are ten multiples of 5: 5, 10, 15, ... , 50, and two of these, 25 and 50 have two 5s. $10 + 2 = 12$

   b) 100! has 24 zeros. There are twenty multiples of 5 (5, 10, 15, ... , 100) and four of these (25, 50, 75, 100) have two 5s. $20 + 4 = 24$

c) 500! has 124 zeros. There are one hundred multiples of 5 $(5, 10, 15, \ldots, 500)$, twenty multiples of $5^2 = 25$, $(25, 50, 75, \ldots, 500)$ and four multiples of $5^3 = 125$, $(125, 250, 375, 500)$. $100 + 20 + 4 = 124$.

2. a) Consider $20! = 20 \cdot 19 \cdot 18 \cdot 17 \cdot 16 \cdot 15 \cdot 14 \cdot 13 \cdot 12 \cdot 11 \cdot 10 \cdot 9 \cdot 8 \cdot 7 \cdot 6 \cdot 5 \cdot 4 \cdot 3 \cdot 2 \cdot 1$

1) Factor out the 10s

$$\left.\begin{array}{l} 20 \to 10 \cdot 2 \\ 15 \to 5 \cdot 3 \\ 10 \to 10 \cdot 1 \\ 5 \to 5 \cdot 1 \\ 2 \to 2 \cdot 1 \end{array}\right\} \text{rewrite these (removing 10s) as 3}$$

2) Then together with the 3 rewrite the one's digits of the remaining terms in the product and group them as shown:

$$\begin{array}{llllllll} 3 \cdot & 9 \cdot & 8 \cdot & 7 \cdot & 6 \cdot & 4 \cdot & 3 \cdot & 2 \cdot \\ & 9 \cdot & 8 \cdot & 7 \cdot & 6 \cdot & 4 \cdot & 3 \\ \hline = \quad 3 \cdot & 81 \cdot & 64 & \cdot & 49 \cdot & 36 \cdot 16 & \cdot & 9 \cdot 2 \\ = & \underbrace{2} & & \cdot & & \underbrace{4} & \cdot & \underbrace{8} \end{array}$$

$2 \times 4 \times 8 = 6\underline{4}$ so the last nonzero digit is 4.

b) Consider $25! = 25 \cdot 24 \cdot 23 \cdot 22 \cdot 21 \cdot 20 \cdot 19 \cdot 18 \cdot 17 \cdot 16 \cdot 15 \cdot 14 \cdot 13 \cdot 12 \cdot$ $11 \cdot 10 \cdot 9 \cdot 8 \cdot 7 \cdot 6 \cdot 5 \cdot 4 \cdot 3 \cdot 2 \cdot 1$

1) Factor out the 10s

$$\left.\begin{array}{l} 25 \to 5 \cdot 5 \\ 20 \to 10 \cdot 2 \\ 15 \to 5 \cdot 3 \\ 10 \to 10 \cdot 1 \\ 5 \to 5 \cdot 1 \\ 8 \to 2 \cdot 2 \cdot 2 \end{array}\right\} \underline{\underline{3}}$$

2) Then together with the 3 rewrite the one's digits of the remaining terms in the product and group them as shown:

$$\underline{3} \cdot 4 \cdot 3 \cdot 2 \cdot 9 \cdot 8 \cdot 7 \cdot 6 \cdot 4 \cdot 3 \cdot 2 \cdot 9 \cdot 7 \cdot 6 \cdot 4 \cdot 3 \cdot 2$$
$$= \quad 3^4 \cdot 4^3 \cdot 2^3 \cdot 9^2 \cdot 8 \cdot 7^2 \cdot 6^2$$
$$\qquad \downarrow \quad \downarrow \quad \downarrow \quad \downarrow \quad \downarrow \quad \downarrow \quad \downarrow$$
$$= \quad 1 \cdot 4 \cdot 8 \cdot 1 \cdot 8 \cdot 9 \cdot 6 = 13{,}824$$

Multiplying the one's digits, the last nonzero digit is 4.

c) Consider $30! = 30 \cdot 29 \cdot 28 \cdot 27 \cdot 26 \cdot 25 \cdot 24 \cdot 23 \cdot 22 \cdot 21 \cdot 20 \cdot 19 \cdot 18 \cdot 17 \cdot$
$16 \cdot 15 \cdot 14 \cdot 13 \cdot 12 \cdot 11 \cdot 10 \cdot 9 \cdot 8 \cdot 7 \cdot 6 \cdot 5 \cdot 4 \cdot 3 \cdot 2 \cdot 1$

1) Factor out the 10s

$$
\left.
\begin{array}{l}
30 \to 3 \cdot 10 \\
25 \to 5 \cdot 5 \\
20 \to 2 \cdot 10 \\
15 \to 3 \cdot 5 \\
10 \to 1 \cdot 10 \\
5 \to 1 \cdot 5 \\
8 \to 2 \cdot 2 \cdot 2
\end{array}
\right\} 3 \cdot 3 = \underline{9}
$$

2) Then together with the 9 rewrite the one's digits of the remaining terms in the product and group them as shown:

$$
\underline{9} \cdot 9 \cdot 8 \cdot 7 \cdot 6 \cdot 4 \cdot 3 \cdot 2 \cdot 9 \cdot 8 \cdot 7 \cdot 6 \cdot 4 \cdot 3 \cdot 2 \cdot 9 \cdot 7 \cdot 6 \cdot 4 \cdot 3 \cdot 2
$$
$$
= 9^4 \cdot 8^2 \cdot 7^3 \cdot 6^3 \cdot 4^3 \cdot 3^3 \cdot 2^3
$$
$$
= 1 \cdot 4 \cdot 3 \cdot 6 \cdot 4 \cdot 7 \cdot 8 = 16{,}128
$$

Multiplying the one's digits, the last nonzero digit is 8.

3.

n	$n!$
1	1
2	2
3	6
4	24
5	120
6	720
7	5040
8	40320
9	362880
10	3628800
11	39916800
12	479001600
13	6227020800
14	87178291200
15	1.30767E+12

On this spreadsheet, $n!$ becomes approximate at $n = 15$. On calculators with 10 digit displays, the answer becomes approximate at $n = 14$.

Some computer applications will give exact values for $n!$ for larger values of n. For example try *MATHEMATICA*.

$25! = 15511210043330985984000000$

$50! = 30414093201713378043612608166064768844377641568960512000000000000$

$100! = 93326215443944152681699238856266700490715968264381621468592963895217599993229915608941463976156518286253697920827223758251185210916864000000000000000000000000$

4. The French mathematician Christian Kramp (1760–1826) of Strasbourg introduced ! for factorial in his book **Elements d'Arithmetique Universelle** in 1808 to avoid printing difficulties.

5. A tic-tac-toe game is played on a 3×3 game board. For the first move there are 9 options.

   For the 2nd, 8; 3rd, 7; 4th, 6; 5th, 5. So there are $9 \cdot 8 \cdot 7 \cdot 6 \cdot 5$ or $15{,}120$ different sequences for the first five moves in the game of tic-tac-toe.

6. a) $25!$

   b) $(12!)(13!)$

   Consider the girls first. There are $12!$ different lines that can be formed. Furthermore, there are $13!$ different lines the boys can form separately. So for each one of the $12!$ lines of girls, $13!$ different lines can be made by adding boys. Thus, there are $(12!)(13!)$ possible lines all together.

Fermat Factorings

1. a) Sum the odd numbers from 1 to 31, and add the result to 185, getting $441 = 21^2$. $185 = (21 + 16)(21 - 16) = 37 \times 5$.

   b) $360 = 2^3 \times 45$. Add the sum of the odd numbers from 1 to 3, to 45, getting $49 = 7^2$. $49 = (7 + 2)(7 - 2) = 9 \times 5$. Therefore $360 = 2^3 \times 3^2 \times 5$.

   c) Sum the odd numbers from 1 to 75, and add the result to 237, getting $1681 = 41^2$. $237 = (41 + 38)(41 - 38) = 79 \times 3$.

   d) $500 = 2^2 \times 125$. Sum the odd numbers from 1 to 19, and add this result to 125, getting $225 = 15^2$. $125 = (15+10)(15-10) = 25 \times 5 = 5^3$. Therefore $500 = 2^2 \times 5^3$.

2. Adding consecutive odd integers to N until the sum is a perfect square can be written as follows: $N + (1 + 3 + 5 + \cdots + (2n - 1)) = k^2$, for some integer k. Since $1 + 3 + 5 + \cdots + (2n - 1) = n^2$, then $N + n^2 = k^2$ and $N = k^2 - n^2 = (k - n)(k + n)$.

3. Consult a book on the history of mathematics, such as Katz, V., **A History of Mathematics**, 3rd edition, NY: Harper Collins, 2008 or read Aczel, A., **Fermat's Last Theorem**, NY: Basic Books, 2007.

Fibonacci Fascinations

For 1, 2 and 3 use the spreadsheet below.

n	Fibonacci n	CS	RD2	RD3	RD5	Ratio L/S	Ratio S/L
1	1	1	1	1	1	—	—
2	1	2	1	1	1	1	1
3	2	4	0	2	2	2	0.5
4	3	7	1	0	3	1.5	0.666666667
5	5	12	1	2	0	1.666666667	0.6
6	8	20	0	2	3	1.6	0.625
7	13	33	1	1	3	1.625	0.615384615
8	21	54	1	0	1	1.615384615	0.619047619
9	34	88	0	1	4	1.619047619	0.617647059
10	55	143	1	1	0	1.617647059	0.618181818
11	89	232	1	2	4	1.618181818	0.617977528
12	144	376	0	0	4	1.617977528	0.618055556
13	233	609	1	2	3	1.618055556	0.618025751
14	377	986	1	2	2	1.618025751	0.618037135
15	610	1596	0	1	0	1.618037135	0.618032787
16	987	2583	1	0	2	1.618032787	0.618034448
17	1597	4180	1	1	2	1.618034448	0.618033813
18	2584	6764	0	1	4	1.618033813	0.618034056
19	4181	10945	1	2	1	1.618034056	0.618033963
20	6765	17710	1	0	0	1.618033963	0.618033999
21	10946	28656	0	2	1	1.618033999	0.618033985
22	17711	46367	1	2	1	1.618033985	0.61803399
23	28657	75024	1	1	2	1.61803399	0.618033988
24	46368	121392	0	0	3	1.618033988	0.618033989
25	75025	196417	1	1	0	1.618033989	0.618033989
26	121393	317810	1	1	3	1.618033989	0.618033989
27	196418	514228	0	2	3	1.618033989	0.618033989
28	317811	832039	1	0	1	1.618033989	0.618033989
29	514229	1346268	1	2	4	1.618033989	0.618033989
30	832040	2178308	0	2	0	1.618033989	0.618033989
31	1346269	3524577	1	1	4	1.618033989	0.618033989
32	2178309	5702886	1	0	4	1.618033989	0.618033989
33	3524578	9227464	0	1	3	1.618033989	0.618033989
34	5702887	14930351	1	1	2	1.618033989	0.618033989
35	9227465	24157816	1	2	0	1.618033989	0.618033989
36	14930352	39088168	0	0	2	1.618033989	0.618033989
37	24157817	63245985	1	2	2	1.618033989	0.618033989
38	39088169	102334154	1	2	4	1.618033989	0.618033989
39	63245986	165580140	0	1	1	1.618033989	0.618033989
40	102334155	267914295	1	0	0	1.618033989	0.618033989

3. a) The cumulative sum of the Fibonacci numbers is $F_1 + F_2 + \cdots + F_n = F_{n+2} - 1$.

   b) The remainder after division by two follows the cyclic pattern of $\underline{1, 1, 0,}$ $\underline{1, 1, 0, \ldots}$

c) The remainder after division by three follows the cyclic pattern of $\underline{1, 1, 2,}$
   $\underline{0, 2, 2, 1, 0,}$ 1, 1, 2, 0, 2, 2, 1, 0, ...

d) The remainder after division by five follows the cyclic pattern of $\underline{1, 1, 2,}$
   $\underline{3, 0, 3, 3, 1, 4, 0, 4, 4, 3, 2, 0, 2, 2, 4, 1, 0,}$ 1, 1, 2, 3, 0, 3, 3, 1, 4, 0, 4, 4, 3,
   2, 0, 2, 2, 4, 1, 0, ...

e) The ratio stabilizes at 1.61803399, and is an approximation of the irrational
   number known as the Golden Ratio. Phidias, a prominent sculptor during
   the Golden Age of Pericles in Ancient Greece (5th century, B.C.) used the
   Golden Ratio extensively. The Golden Ratio to 25 decimal positions is
   1.6180339887498948482045868.

f) The ratio stabilizes at 0.61803399, an approximation of the inverse of the
   Golden Ratio.

4. Consult history of mathematics books and Posamentier, A.S. and Lehmann, I.,
   The Fabulous Fibonacci Numbers, NY: Prometheus Books, 2007.

5. Consult geometry texts and also Vajda, S., **Fibonacci and Lucas Numbers and
   the Golden Section: Theory and Applications,** NY: Dover, 2008.

Footsteps of Lagrange

1.

n	Ways to Write as Sum of at Most 4 Squares			s_n
1	1^2			1
2	$1^2 + 1^2$			1
3	$1^2 + 1^2 + 1^2$			1
4	2^2;	$1^2 + 1^2 + 1^2 + 1^2$		2
5	$2^2 + 1^2$			1
6	$2^2 + 1^2 + 1^2$			1
7	$2^2 + 1^2 + 1^2 + 1^2$			1
8	$2^2 + 2^2$			1
9	3^2;	$2^2 + 2^2 + 1^2$		2
10	$3^2 + 1^2$;	$2^2 + 2^2 + 1^2 + 1^2$		2
11	$3^2 + 1^2 + 1^2$			1
12	$3^2 + 1^2 + 1^2 + 1^2$	$2^2 + 2^2 + 1^2$		2
13	$3^2 + 2^2$;	$2^2 + 2^2 + 2^2 + 1$		2
14	$3^2 + 2^2 + 1$			1
15	$3^2 + 2^2 + 1^2 + 1^2$			1
16	4^2;	$2^2 + 2^2 + 2^2 + 2^2$		2
17	$4^2 + 1^2$;	$3^2 + 2^2 + 2^2$		2
18	$4^2 + 1^2 + 1^2$;	$3^2 + 3^2$;	$3^2 + 2^2 + 2^2 + 1^2$	3
19	$4^2 + 1^2 + 1^2 + 1^2$;	$3^2 + 3^2 + 1^2$		2
20	$4^2 + 2^2$;	$3^2 + 3^2 + 1^2 + 1^2$		2

2. Answers will vary. For example, in the table, square numbers greater than 1 can
   be written in two ways.

3. Answers will vary. $27 = 5^2 + 1^2 + 1^2 = 4^2 + 3^2 + 1^2 + 1^2 = 3^2 + 3^2 + 3^2$.
4. Answers will vary. $45 = 6^2 + 3^2 = 6^2 + 2^2 + 2^2 + 1 = 5^2 + 4^2 + 2^2 = 4^2 + 4^2 + 3^2 + 2^2$.

Geoboard Journeys

1.

a) 4 x 4
    5 paths

b) 5 x 5
    14 paths

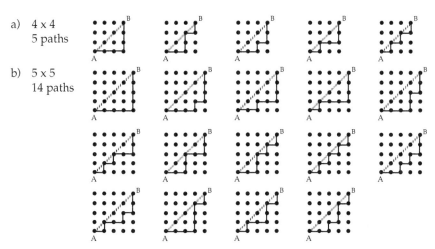

2. 1, 2, 5, 14 are Catalan numbers.
3. In the grid below, the numbers represent the number of paths to that marked point from point A using the given rules.

   a) 2 × 2   b) 3 × 3   c) 4 × 4   d) 5 × 5
      2 paths   6 paths   20 paths   70 paths

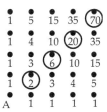

4. The numbers in the grid are those in Pascal's triangle. The circled numbers are the middle numbers in the even rows of the triangle. The numbers in 2 can be found by dividing the circled number by one-half its row number plus one. That is $2/2 = 1, 6/3 = 2, 20/4 = 5, 70/5 = 14$.
5. Answers will vary.

Hexagons in Black and White

1.

n	Number of Black Chips	Number of White Chips	Total Chips
0	1	0	$1 + 0 \times 6 = 1$
1	1	6	$1 + 1 \times 6 = 7$
2	$1 + 12$	6	$7 + 2 \times 6 = 19$
3	$1 + 12$	$6 + 18$	$19 + 3 \times 6 = 37$
4	$1 + 12 + 24$	$6 + 18$	$37 + 4 \times 6 = 61$
5	$1 + 12 + 24$	$6 + 18 + 30$	$61 + 5 \times 6 = 91$
6	$1 + 12 + 24 + 36$	$6 + 18 + 30$	$91 + 6 \times 6 = 127$
7	$1 + 12 + 24 + 36$	$6 + 18 + 30 + 42$	$127 + 7 \times 6 = 169$
8	$1 + 12 + 24 + 36 + 48$	$6 + 18 + 30 + 42$	$169 + 8 \times 6 = 217$
9	$1 + 12 + 24 + 36 + 48$	$6 + 18 + 30 + 42 + 54$	$217 + 9 \times 6 = 271$
10	$1 + 12 + 24 + 36 + 48 + 60$	$6 + 18 + 30 + 42 + 54$	$271 + 10 \times 6 = 331$

2. a) $C_0 = 1, C_n = C_{n-1} + 6n, n \geq 1$
   b) $f(n) = 3n^2 - 3n + 1$
3. Answers will vary.

Highly Composite Numbers

1. a) 2, 4, 6, 12, 24
   b) 2, 4, 6, 12, 24, 36, 48
2. a) $60 = 2^2 \cdot 3 \cdot 5$ b) $144 = 2^4 \cdot 3^2$ c) $512 = 2^9$

$N_{div} = (2+1)(1+1)(1+1)$ $N_{div} = (4+1)(2+1)$ $N_{div} = (9+1)$
$= (3)(2)(2) = 12$ $= (5)(3) = 15$ $= 10$

3. a) $100 = 2^2 \cdot 5^2$ b) $180 = 2^2 \cdot 3^2 \cdot 5$ c) $250 = 2 \cdot 5^3$

$N_{div} = (2+1)(2+1)$ $N_{div} = (2+1)(2+1)(1+1)$ $N_{div} = (1+1)(3+1)$
$= 9$ $= 18$ $= 8$

not highly composite   highly composite   not highly composite
(36 has 9 divisors)   (24 has 8 divisors)

4. $1000 = 2^3 \cdot 5^3$ $120 = 2^3 \cdot 3 \cdot 5$ is the smallest highly
   $N_{div} = (3+1)(3+1) = 16$ composite number that has the same number
   of divisors as 1000.
5. Try Kanigel, R., **The Man Who Knew Infinity: A Life of the Genius Ramanujan**, NY: Washington Square Press (paperback), 1992.

Honest Number Hunt

1. No. For example, there do not appear to be Honest numbers in the Indonesian, French, Polish, and Swahili languages.

A sample Honest Number Chart.

Chinese (Zhongguo)	er (2)	san (3)
English	four (4)	
Hawaiian	alima (5)	umikumamalua (12)
	umikumamakolu (13)	
Japanese	ni (2)	san (3)
	muttsu (6)	nanatsu (7)
	kokonotsu (9)	
Mende (West Africa)	nani (4)	lorlu (5)
	worfila (7)	puu mahu yila (11)
Russian*	nhb (3)	jlbyyflwfnm (11)
Sanskrit	panca (5)	
Spanish	cinco (5)	
Tagalog (Philippines)	apat (4)	
Zula	hlanu (5)	

*In Russian, the symbol 'b' at the end of certain words is included as a letter.

2. Answers will vary.

A Juggling Act

Positive multiples represent filling that many times, and negative multiples represent emptying that many times.

1. $1 = 3 \cdot 5 - 2 \cdot 7$

5	7
0	0
5	0
0	5
5	5
3	7
3	0
0	3
5	3
1	7
1	0

2. $1 = 2 \cdot 4 - 1 \cdot 7$

4	7
0	0
4	0
0	4
4	4
1	7
1	0

3. $1 = 2 \cdot 8 - 3 \cdot 5$

5	8
0	0
0	8
5	3
0	3
3	0
3	8
5	6
0	6
5	1
0	1

4. $1 = 2 \cdot 8 - 3 \cdot 5$ See problem 3.

$2 = 2 \cdot 5 - 1 \cdot 8$	$8 = 1 \cdot 8 - 0 \cdot 5$
$3 = 1 \cdot 8 - 1 \cdot 5$	$9 = 3 \cdot 8 - 3 \cdot 5$
$4 = 4 \cdot 5 - 2 \cdot 8$	$10 = 5 \cdot 8 - 6 \cdot 5$
$5 = 1 \cdot 5 - 0 \cdot 8$	$11 = 2 \cdot 8 - 1 \cdot 5$
$6 = 2 \cdot 8 - 2 \cdot 5$	$12 = 4 \cdot 5 - 1 \cdot 8$
$7 = 3 \cdot 5 - 1 \cdot 8$	$13 = 1 \cdot 8 + 1 \cdot 5$

5. No. No combination of positive and negative multiples of 6 and 8 equals 1.
6. a, b must be relatively prime.
7. _____

8	5	3
0	0	0
3	5	0
3	2	3
6	2	0
6	0	2
1	5	2
1	4	3
4	4	0

Marble Art

This is a modeling activity. Encourage students to create their own designs.

A Medieval Pattern

1. $s_5 = 103$. See the figure to verify and confirm the pattern.
2.

n	1	2	3	4	5	6	7	8	9	10
Unit Square Count s_n	7	19	39	67	103	147	199	259	327	403

3. The starting number is an odd number. The number of squares added on at each new step is 4 times an odd number, which is an even number. The sum of an odd number and an even number is an odd number.

4. <u>Recursive formula</u>:

$$s_1 = 7, \qquad s_2 = 19 = s_1 + 12 = s_1 + 4 \cdot 3$$
$$s_3 = 39 = s_2 + 20 = s_2 + 4 \cdot 5$$
$$s_4 = 67 = s_3 + 28 = s_3 + 4 \cdot 7$$
$$s_1 = 7, \qquad s_n = s_{n-1} + 4(2n - 1), \quad n = 2, 3, 4, \dots$$

5. <u>Explicit formula</u>:

$$s_1 = 7 = 4 + 3 = 4 \cdot 1^2 + 3$$
$$s_2 = 19 = 16 + 3 = 4 \cdot 2^2 + 3$$
$$s_3 = 39 = 36 + 3 = 4 \cdot 3^2 + 3$$
$$s_n = 4n^2 + 3, \quad n = 1, 2, 3, \dots$$

6. Answers will vary.

Mysterious Mountains and Binary Trees

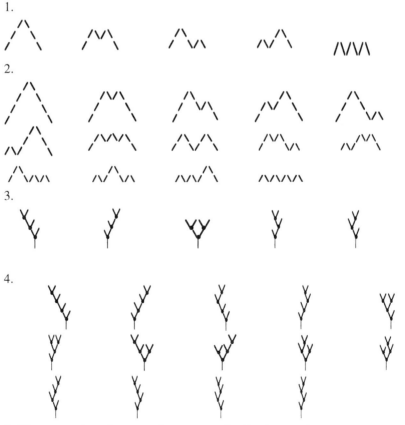

1.

2.

3.

4.

5. The mountain and tree results appear to be Catalan numbers.

Pentagonal Play

1. a) $P_6 = 51$ $\quad\quad\quad P_7 = 70$ $\quad\quad P_8 = 92$ $\quad\quad P_9 = 117$ $\quad\quad P_{10} = 145$

   b) $P_n = \dfrac{3n^2 - n}{2}$

2. $P_8 = 64 + 28 = S_8 + T_7$ $\quad\quad P_9 = 81 + 36 = S_9 + T_8$

   $P_{10} = 100 + 45 = S_{10} + T_9$ $\quad P_n = S_n + T_{n-1} = n^2 + \dfrac{(n-1)n}{2}$

3. $P_4 = 22 = 10 + 6 + 6 = T_4 + T_3 + T_3$

   $P_5 = 35 = 15 + 10 + 10 = T_5 + T_4 + T_4$

   $P_6 = 51 = 21 + 15 + 15 = T_6 + T_5 + T_5$

   $P_7 = 70 = 28 + 21 + 21 = T_7 + T_6 + T_6$

   $P_8 = 92 = 36 + 28 + 28 = T_8 + T_7 + T_7$

   $P_9 = 117 = 45 + 36 + 36 = T_9 + T_8 + T_8$

   $P_{10} = 145 = 55 + 45 + 45 = T_{10} + T_9 + T_9$

   In general, $P_n = T_n + 2T_{n-1} = \dfrac{n(n+1)}{2} + 2\left[\dfrac{n(n-1)}{2}\right]$

   $$= \dfrac{n^2 + n + 2(n^2 - n)}{2} = \dfrac{3n^2 - n}{2}$$

4.

n	1	2	3	4	5	6	7	8	9	10
P_n	1	5	12	22	35	51	70	92	117	145
Sum of Digits	1	5	3	4	8	6	7	2	9	1
n	11	12	13	14	15	16	17	18	19	20
P_n	176	210	247	287	330	376	425	477	532	590
Sum of Digits	5	3	4	8	6	7	2	9	1	5

The digits 1 through 9 appear in the digital sums.

The repetition pattern for the digital sum of the pentagonal numbers is 1, 5, 3, 4, 8, 6, 7, 2, 9.

5. 3, 4, and 8 do not occur as unit's digits. The unit's digit in pentagonal numbers goes through a 20-digit cycle.

   1 5 2 2 5 1 $\quad\quad$ 0 2 7 5 6 0 7 7 0 6 5 7 2 0

   Observe the mirror pattern within each group.

Perfect Number Patterns

1. For $n = 2, 3, 5, 7$ the formula $2^{n-1}(2^n - 1)$ gives the perfect numbers 6, 28, 496, 8128.

2. a) $\log(2^{88} \times 2^{89}) = 177 \log 2 = \underline{53.2823}$ $\quad\quad$ Number of digits: 54

   b) $\log(2^{1278} \times 2^{1279})$ $\quad\quad\quad\quad\quad\quad\quad\quad$ Number of digits: 770

   $\quad\quad = 2557 \log 2 = \underline{769.73371}$

   c) $\log(2^{4422} \times 2^{4423}) = 8845 \log 2$ $\quad\quad$ Number of digits: 2663

   $\quad\quad = \underline{2662.61035}$

d) $\log(2^{21700} \times 2^{21701})$ Number of digits: 13066
   $= 43401 \log 2 = \underline{13065}.003$
e) $\log(2^{132048} \times 2^{132049})$ Number of digits: 79502
   $= 264097 \log 2 = \underline{79501}.1199$

3. $496 = 256 + 128 + 64 + 32 + 16$ $8128 = 4096 + 2048 + 1024 + 512$
   $+ 256 + 128 + 64$

4. Using substitution $2^{n-1}(2^n - 1) = 2^{n-1}(1 + 2 + 2^2 + \cdots + 2^{n-1}) = 2^{n-1} + 2^n + 2^{n+1} + \cdots + 2^{2n-2}$ which are all powers of 2.

5. An excellent comprehensive Internet resource is primes.utm.edu.
   A variety of number theory texts and various works by Paulo Ribenboim are helpful.

Prime Magic

1. Answers will vary, but these are possibilities found by trial and error checking with arithmetic sequences.

2 Primes	3 Primes	4 Primes	5 Primes	6 Primes	7 Primes	8 Primes
83 84, 85, 86, 87, 88, **89**, 90, 91	**23**, 24, 25, 26, 27, 28, **29**, 30, **31**	1, **2**, **3**, 4, **5**, 6, **7**, 8, 9	**2**, **5**, 8, **11**, 14, **17**, 20, **23**, 26	**7**, **13**, **19**, 25, **31**, **37**, **43**, 49, 55	**3**, **5**, **7**, 9, **11**, **13**, 15, **17**, **19**	**5**, **11**, **17**, **23**, 29, 35, **41**, **47**, **53**

84	89	88
91	87	83
86	85	90

M. C.=261

24	29	28
31	27	23
26	25	30

M. C.=81

2	7	6
9	5	1
4	3	8

M. C.=15

5	20	17
26	14	2
11	8	23

M. C.=42

13	43	37
55	31	7
25	19	49

M. C.=93

5	15	13
19	11	3
9	7	17

M. C.=33

11	41	35
53	29	5
23	17	47

M. C.=87

2. Investigate the topic of magic squares in recreational mathematics materials online and in books such as Pickover, C.A., **The Zen of Magic Squares, Circles, and Stars: An Exhibition of Surprising Structures Across Dimensions**, NJ: Princeton University Press, 2002.

Pythagorean Triple Pursuits

1. a)

n	a $2n+1$	b $2n^2+2n$	c $2n^2+2n+1$	n	a $2n+1$	b $2n^2+2n$	c $2n^2+2n+1$
1	3	4	5	13	27	364	365
2	5	12	13	14	29	420	421
3	7	24	25	15	31	480	481
4	9	40	41	16	33	544	545
5	11	60	61	17	35	612	613
6	13	84	85	18	37	684	685
7	15	112	113	19	39	760	761
8	17	144	145	20	41	840	841
9	19	180	181	21	43	924	925
10	21	220	221	22	45	1012	1013
11	23	264	265	23	47	1104	1105
12	25	312	313	24	49	1200	1201

n	a $2n+1$	b $2n^2+2n$	c $2n^2+2n+1$	n	a $2n+1$	b $2n^2+2n$	c $2n^2+2n+1$
25	51	1300	1301	38	77	2964	2965
26	53	1404	1405	39	79	3120	3121
27	55	1512	1513	40	81	3280	3281
28	57	1624	1625	41	83	3444	3445
29	59	1740	1741	42	85	3612	3613
30	61	1860	1861	43	87	3784	3785
31	63	1984	1985	44	89	3960	3961
32	65	2112	2113	45	91	4140	4141
33	67	2244	2245	46	93	4324	4325
34	69	2380	2381	47	95	4512	4513
35	71	2520	2521	48	97	4704	4705
36	73	2664	2665	49	99	4900	4901
37	75	2812	2813	50	101	5100	5101

b) $a^2 + b^2 = (2n+1)^2 + (2n^2+2n)^2 = 4n^2 + 4n + 1 + 4n^4 + 8n^3 + 4n^2$
$$= 4n^4 + 8n^3 + 8n^2 + 4n + 1$$
and $c^2 = (2n^2+2n+1)^2 = 4n^4 + 8n^3 + 8n^2 + 4n + 1$
So, $a^2 + b^2 = c^2$ for all n.

2. a) $a = 2mn, \quad b = m^2 - n^2, \quad c = m^2 + n^2$

n	m $=n+1$	a	b	c	m $=n+2$	a	b	c	m $=n+3$	a	b	c
1	2	4	3	5	3	6	8	10	4	8	15	17
2	3	12	5	13	4	16	12	20	5	20	21	29
3	4	24	7	25	5	30	16	34	6	36	27	45
4	5	40	9	41	6	48	20	52	7	56	33	65
5	6	60	11	61	7	70	24	74	8	80	39	89
6	7	84	13	85	8	96	28	100	9	108	45	117
7	8	112	15	113	9	126	32	130	10	140	51	149
8	9	144	17	145	10	160	36	164	11	176	57	185
9	10	180	19	181	11	198	40	202	12	216	63	225
10	11	220	21	221	12	240	44	244	13	260	69	269
$\vdots$	$\vdots$	$\vdots$	$\vdots$	$\vdots$	$\vdots$	$\vdots$	$\vdots$	$\vdots$	$\vdots$	$\vdots$	$\vdots$	$\vdots$
50	51	5100	101	5101	52	5200	204	5204	53	5300	309	5309

b) $a^2 + b^2 = (2mn)^2 + (m^2 - n^2)^2 = 4m^2n^2 + m^4 - 2m^2n^2 + n^4$
$$= m^4 + 2m^2n^2 + n^4$$
and $c^2 = (m^2 + n^2)^2 = m^4 + 2m^2n^2 + n^4$
So, $a^2 + b^2 = c^2$ for all m and n, $m > n$

3. a) There are numerous resources that cover this topic. Answers may vary in wording, but a primitive Pythagorean triple is a Pythagorean triple in which the three numbers have only a common factor of 1.

b) Reports should include the following: various formulas for finding Primitive Pythagorean triples include those attributed to
Pythagoras:
$a = 2n + 1, \quad b = 2n^2 + 2n, \quad c = 2n^2 + 2n + 1; \quad n = 1, 2, 3, \ldots$

Ancients:
$$a = r, \quad b = \frac{r^2 - 1}{2}, \quad c = \frac{r^2 + 1}{2}; \quad r = 3, 5, 7, 9, \ldots$$
Plato:
$$a = 2m, \quad b = m^2 - 1, \quad c = m^2 + 1; \quad m = 2, 3, 4, 5, \ldots$$
Euclid:
$$a = 2st, \quad b = s^2 - t^2, \quad c = s^2 + t^2; \quad s > t, s, t = 1, 2, 3, \ldots$$
Moderns:
$$a = mn, \quad b = \frac{m^2 - n^2}{2}, \quad c = \frac{m^2 + n^2}{2}; \quad m > n, m, n \text{ are relatively}$$
$$\text{prime}, m, n = 1, 3, 5, 7, \ldots$$

4.

3, 4, 5	24, 32, 40	45, 60, 75	10, 24, 26
6, 8, 10	27, 36, 45	48, 64, 80	15, 36, 39
9, 12, 15	30, 40, 50	51, 68, 85	20, 48, 52
12, 16, 20	33, 44, 55	54, 72, 90	25, 60, 65
15, 20, 25	36, 48, 60	57, 76, 95	30, 72, 78
18, 24, 30	39, 52, 65	**5, 12, 13**	35, 84, 91
21, 28, 35	42, 56, 70		

7, 24, 25	32, 60, 68	**20, 21, 29**	**33, 56, 65**
14, 48, 50	40, 75, 85	40, 42, 58	**48, 55, 73**
21, 72,75	**9, 40, 41**	60, 63, 87	**13, 84, 85**
8, 15, 17	18, 80, 82	**28, 45, 53**	**36, 77, 85**
16, 30, 34	**12, 35, 37**	**11, 60, 61**	**39, 80, 89**
24, 45, 51	24, 70, 74	**16, 63, 65**	**65, 72, 97**

Primitive Pythagorean triples are boxed.

Seeking Honesty in Numbers

1. 63: sixty-three → 10: ten → 3: three → 5: five → 4: four
   The number trail has 5 numbers: 63→ 10 → 3 → 5 → 4
2. 163: one hundred sixty-three → 20: twenty → 6: six → 3: three → 5: five →
   4: four
   The number trail has 6 numbers: 163 → 20 → 6 → 3 → 5 → 4
3. 999: nine hundred ninety-nine → 21: twenty-one → 9: nine → 4: four
   The number trail has 4 numbers: 999 → 21 → 9 → 4
4. 8999: eight thousand nine hundred ninety-nine → 34: thirty-four → 10: ten
   →3: three → 5: five → 4: four
   The number trail has 6 numbers: 8999 → 34 → 10 → 3 → 5 → 4

The answers to questions 5–8 will vary depending on the particular number chosen.
One example of each is given.

5. 12467: twelve thousand four hundred sixty-seven →35: thirty-five → 10: ten
   → 3: three → 5: five → 4: four
   The number trail has 6 numbers: 12,467 → 35 → 10 → 3 → 5 → 4
6. 742683: seven hundred forty-two thousand six hundred eighty-three →49:
   forty-nine → 9: nine → 4: four

The number trail has 4 numbers: $742,683 \rightarrow 49 \rightarrow 9 \rightarrow 4$

7. 5824391: five million eight hundred twenty-four thousand three hundred ninety-one $\rightarrow$ 62: sixty-two $\rightarrow$ 8: eight $\rightarrow$ 5: five $\rightarrow$ 4: four

The number trail has 5 numbers: $5,824,391 \rightarrow 62 \rightarrow 8 \rightarrow 5 \rightarrow 4$

8. 38690214: thirty-eight million six hundred ninety thousand two hundred fourteen $\rightarrow$ 60: sixty $\rightarrow$ 5: five $\rightarrow$ 4: four

The number trail has 4 numbers: $38,690,214 \rightarrow 60 \rightarrow 5 \rightarrow 4$

The Super Sum

1. Add the first and last, second and second last, and so on. The sum is 101 in each case. There are 50 pairs so the total is $50 \times 101 = 5050$. Encourage discussion on the ways to achieve the result.

2. a) $1 + 99 = 100$, $2 + 98 = 100$, and so on. There are 49 pairs with sum 100 and $49 \times 100 = 4900$. 50 has no partner, so the total is $4900 + 50 = 4950$. This problem may also be solved by subtracting 100 from the answer in (1). $5050 - 100 = 4950$.

   b) $1 + 101 = 102$, $2 + 100 = 102$, and so on. There are 50 pairs with sum 102 and $50 \times 102 = 5100$. 51 has no partner, so the total is $5100 + 51 = 5151$. This problem may also be solved by adding 101 to the answer in (1). $5050 + 101 = 5151$.

   c) $1 + 500 = 501$, $2 + 499 = 501$, and so on. There are 250 pairs with sum 501 and $250 \times 501 = 125,250$.

   d) $1 + 175 = 176$, $2 + 174 = 176$, and so on. There are 80 pairs with sum 176 and $80 \times 176 = 15,312$. Since 88 has no partner, the total is $15,312 + 88 = 15,400$.

3. a) Total is $\frac{n}{2} \cdot (n + 1)$ or $\frac{n(n + 1)}{2}$.

   b) Total is $\left(\frac{n - 1}{2} \cdot (n + 1) \right) + \frac{n + 1}{2}$ which simplifies to $\frac{n(n + 1)}{2}$.

4. Total for any n is $\frac{n(n + 1)}{2}$

5. a) $2 + 4 + 6 + \cdots + 200 = 2(1 + 2 + \cdots + 100) = 2 \cdot \frac{100(101)}{2} = 10,100$.

   b) $3 + 6 + 9 + \cdots + 255 = 3(1 + 2 + \cdots + 85) = 3 \cdot \frac{85(86)}{2} = 10,965$.

6. a) 168 rectangles.

			length					
	$\times$	1	2	3	4	5	6	7
	1	$21 + 18 + 15 + 12 + 9 + 6 + 3$						
		$3(7 + 6 + 5 + 4 + 3 + 2 + 1) = 84$						
width	2	$14 + 12 + 10 + 8 + 6 + 4 + 2$						
		$2(7 + 6 + 5 + 4 + 3 + 2 + 1) = 56$						
	3	$1(7 + 6 + 5 + 4 + 3 + 2 + 1) = 28$						
$(1 + 2 + 3)(1 + 2 + 3 + 4 + 5 + 6 + 7) = 168$								

length

×	1	2	3	4	5	6	7	8

width

	1	$32 + 28 + 24 + 20 + 16 + 12 + 8 + 4$
		$4(8 + 7 + 6 + 5 + 4 + 3 + 2 + 1) = 144$
	2	$24 + 21 + 18 + 15 + 12 + 9 + 6 + 3$
		$3(8 + 7 + 6 + 5 + 4 + 3 + 2 + 1) = 108$
	3	$16 + 14 + 12 + 10 + 8 + 6 + 4 + 2$
		$2(8 + 7 + 6 + 5 + 4 + 3 + 2 + 1) = 72$
	4	$1(8 + 7 + 6 + 5 + 4 + 3 + 2 + 1) = 36$

$(1 + 2 + 3 + 4)(1 + 2 + 3 + 4 + 5 + 6 + 7 + 8) = 360$

7. a) Using the pattern in problem 6, the rectangle count for a $5 \times n$ rectangle is

$$(1 + 2 + 3 + 4 + 5)(1 + 2 + \cdots + n) = 15 \cdot \frac{n(n+1)}{2}$$

b) The rectangle count for an $m \times n$ rectangle is $\dfrac{m(m+1)}{2} \cdot \dfrac{n(n+1)}{2}$

The Tower of Hanoi and the Reve's Puzzle

1.

Number of Disks n	1	2	3	4	5	6	7	8
Least # of Moves M_n	1	3	7	15	31	63	127	255

Number of Disks n	9	10	n
Least # of Moves M_n	511	1023	$2^n - 1$

2. a) To move 4 disks first move 3 so the 4th can be placed on another spike and then move 3 back onto the 4th. Total moves $M_4 = M_3 + 1 + M_3 = 2M_3 + 1$.
   b) Similarly, $M_{10} = M_9 + 1 + M_9 = 2M_9 + 1$
   c) $M_n = M_{n-1} + 1 + M_{n-1} = 2M_{n-1} + 1$

3. a) about 17 minutes   c) about 12 days   e) about 584,942,417,400 years
   b) about 9 hours   d) about 34 years

4.

Number of Cheeses n	1	2	3	4	5	6	7	8
Least # of Moves S_n	1	3	5	9	13	17	25	33

5. It appears that the result for 4 cheeses can be expressed in terms of the result for 3 cheeses ($9 = S_4 = S_3 + 2^2 = 5 + 2^2$), and the result for 5 cheeses can be expressed in terms of the result for 4 cheeses ($13 = S_5 = S_4 + 2^2 = 9 + 2^2$). However, the generalization for S_n is not as apparent as the one for the Tower of Hanoi problem.

6. 10 cheeses require 49 moves, and 21 cheeses require 321 moves.

Triangular Number Turnarounds

1. Since $T_n = \dfrac{n(n+1)}{2}$, look for the smallest n such that $n(n+1) > 200$. By trial and error, $n = 14$. So $T_{14} = 105$, $T_{15} = 120$, and $T_{16} = 136$ are the first three triangular numbers after 100.

2. Using trial and error, $T_{20} = 210$.

3.

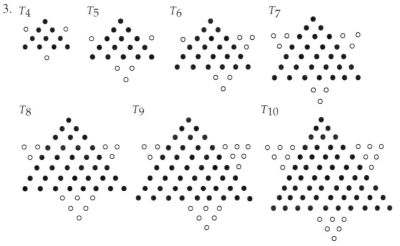

n		Minimum dot moves	
1	0	$0+0+0$	$3T_0$
2	1	$0+0+1$	$2T_0 + T_1$
3	2	$0+1+1$	$T_0 + 2T_1$
4	3	$1+1+1$	$3T_1$
5	5	$1+1+3$	$2T_1 + T_2$
6	7	$1+3+3$	$T_1 + 2T_1$
7	9	$3+3+3$	$3T_2$
8	12	$3+3+6$	$2T_2 + T_3$
9	15	$3+6+6$	$T_2 + 2T_3$
10	18	$6+6+6$	$3T_3$

4. The spreadsheet illustrating the first 50 triangular numbers will show that there are 24 evens and 26 odds. Extending the spreadsheet through 100 will show that the number of even and odd triangular numbers is equal. There are 50 of each.

Trying Trapezoids

1. a) 9, 12, 14   b) 15, 21, 26, 30, 33, 35
   c) $T_k - T_1, T_k - T_2, T_k - T_3, \ldots, T_k - T_{k-2}$

2. a) 14, 18   b) 20, 25, 30   c) $3k - 3, 3k, 3k + 3; k > 3$
3. a) 11   b) 27
      2 rows: $5 + 6$ 2 rows: $13 + 14$
                                                    3 rows: $8 + 9 + 10$
                                                    6 rows: $2 + 3 + 4 + 5 + 6 + 7$

   c) 33
      2 rows: $16 + 17$
      3 rows: $10 + 11 + 12$
      6 rows: $3 + 4 + 5 + 6 + 7 + 8$

4. Let r be the number of rows, f be the number of dots in first row, l be the number of dots in last row, TZ be the trapezoidal number; then $TZ = \dfrac{r}{2}(f + l)$.

5. a) $2 + 3 + 4 + 5$ c) $2 + 3 + 4 + \cdots + 10$
      $= (2 \times 4) + T_3 = 8 + 6$ $= (2 \times 9) + T_8 = 18 + 36$
   b) $2 + 3 + 4 + 5 + 6$ d) $2 + 3 + 4 + \cdots + 20$
      $= (2 \times 5) + T_4 = 10 + 10$ $= (2 \times 19) + T_{18} = 38 + 171$

6. a) and b) Sketch the figure to verify.
   c) 10 hexagons, 20 triangles   d) 15 hexagons, 30 triangles
   e) 45 hexagons, 90 triangles

Solutions to Exercises

In addition to solutions to the exercises, some references are provided. Some of the references are for general reading on related material. Others are more scholarly references for the researcher to pursue.

EXERCISE 1

    1. $\{1, 2, 4, 8\}$ **2.** $\{1, 2, 5, 10\}$ **3.** $\{1, 2, 4, 5, 10, 20\}$ **4.** $\{1, 5\}$

    5. $\{1, 29\}$ **6.** $\{1, 2, 4, 5, 8, 10, 20\}$ **7.** $\{1, 41\}$ **8.** $\{1\}$ **9.** $\{1\}$

EXERCISE 2

    1. a) $\{1, 2, 3, 6, 9, 18\}$, $\{1, 2, 3, 6, 9\}$, composite

        b) $\{1, 23\}$, $\{1\}$, prime

        c) $\{1, 2, 3, 4, 6, 8, 12, 24\}$, $\{1, 2, 3, 4, 6, 8, 12\}$, composite

        d) $\{1, 2, 4, 7, 14, 28\}$, $\{1, 2, 4, 7, 14\}$, composite

        e) $\{1, 31\}$, $\{1\}$, prime

        f) $\{1, 37\}$, $\{1\}$, prime

        g) $\{1, 3, 13, 39\}$, $\{1, 3, 13\}$, composite

        h) $\{1, 59\}$, $\{1\}$, prime

    2. a) 1, 3, 7, 9   b) 0, 1, 2, 3, 4, 5, 6, 7, 8, 9

    Not all integers that end in 1, 3, 7, 9 are primes, but all primes (except 2 and 5) do end in 1, 3, 7, or 9.

    3. The next prime after 7 is 11 and the first multiple of 11 that is not already crossed out is 11×11, which is greater than 100. Thus, all primes less than 100 have been identified once the multiples of 2, 3, 5, and 7 have been removed.

    4. a) The last prime whose multiples are removed is 13. The next number not already crossed out is 17×17 which is larger than 200.

        b) The last prime whose multiples are removed is 17. The next number not already crossed out is 19×19 which is larger than 300.

EXERCISE 3

1. prime   **2.** composite, $2^2 \times 3 \times 7$ **3.** composite, 3×31
4. prime   **5.** composite, $2 \times 3 \times 19$ **6.** prime
7. composite, 7×19 **8.** composite, $2^4 \times 3^2$ **9.** composite, $2^3 \times 19$
10. composite, $2^3 \times 5^2$ **11.** prime   **12.** composite, $3^2 \times 5 \times 11$

EXERCISE 4

1. composite, $2 \times 3 \times 5$ **2.** prime   **3.** composite, $2^5 \times 3$
4. composite, 3×37 **5.** composite, $2^2 \times 31$ **6.** composite, $2^2 \times 7^2$
7. prime   **8.** composite, 17^2 **9.** composite, 17×19
10. composite, 7^3 **11.** composite, 7×71 **12.** composite, 19×29

EXERCISE 5

1. $2^0, 2^1, 3^1, 2^2, 2^1 \times 3^1, 2^3, 2^2 \times 3^1, 2^3 \times 3^1$

2. $2^0, 2^1, 3^1, 2^2, 3^2, 2^1 \times 3^1, 2^3, 2^2 \times 3^1, 2^1 \times 3^2, 2^3 \times 3^1, 2^2 \times 3^2, 2^3 \times 3^2$

3. $2^0, 2^1, 3^1, 2^2, 3^2, 2^1 \times 3^1, 2^3, 3^3, 2^2 \times 3^1, 2^1 \times 3^2, 2^3 \times 3^1, 2^2 \times 3^2,$
$2^1 \times 3^3, 2^3 \times 3^2, 2^2 \times 3^3, 2^3 \times 3^3$

4. $2^0, 2^1, 3^1, 5^1, 2^1 \times 3^1, 2^1 \times 5^1, 3^1 \times 5^1, 2^1 \times 3^1 \times 5^1$

5. $2^0, 2^1, 3^1, 5^1, 2^2, 2^1 \times 3^1, 2^1 \times 5^1, 3^1 \times 5^1, 2^2 \times 3^1, 2^2 \times 5^1, 2^1 \times 3^1 \times 5^1,$
$2^2 \times 3^1 \times 5^1$

6. $2^0, 2^1, 3^1, 5^1, 2^2, 2^1 \times 3^1, 2^1 \times 5^1, 3^1 \times 5^1, 3^2, 2^1 \times 3^1 \times 5^1, 2^2 \times 3^1,$
$2^2 \times 5^1, 2^1 \times 3^2, 3^2 \times 5^1, 2^2 \times 3^2, 2^2 \times 3^1 \times 5^1, 3^2 \times 2^1 \times 5^1, 2^2 \times 3^2 \times 5^1$

7. 4

8. 6

9. 9

10. 12

11. $p+1$

12. $2(n+1)$

13. $(p+1)(q+1)$

EXERCISE 6

1. $54 = 2 \times 3^3$, $60 = 2^2 \times 3 \times 5$, gcd $= 2 \times 3 = 6$, lcm $= 2^2 \times 3^3 \times 5 = 540$;
$\dfrac{54 \times 60}{6} = 540$

2. $40 = 2^3 \times 5$, $28 = 2^2 \times 7$, gcd $= 2^2 = 4$; $\dfrac{40 \times 28}{4} = 280 = $ lcm

3. $36 = 2^2 \times 3^2$, $123 = 3 \times 41$, lcm $= 2^2 \times 3^2 \times 41 = 1476$;
$1476 = \dfrac{36 \times 123}{\text{gcd}}$, gcd $= 3$

4. $(1, 540)$ $(2, 270)$ $(3, 180)$ $(4, 135)$ $(5, 108)$ $(6, 90)$ $(9, 60)$ $(10, 54)$ $(12, 45)$
$(15, 36)$ $(18, 30)$ $(20, 27)$

5. $P = a \times b$ is the product of the prime factorizations of each number.
$G = \gcd(a, b)$ is the common part in the prime factorizations. Once P is divided by G the part that remains is, by definition, the lcm.

EXERCISE 7

1. $a, b, c, f, g, h,$ and l are relatively prime.
2. Yes. Any two primes have only 1 as a common divisor.
3. Yes. Consecutive numbers only differ by one and do not have any common factors other than 1.
4. $\phi(1) = 1: 1$ $\phi(2) = 1: 1$ $\phi(3) = 2: 1, 2$
$\phi(4) = 2: 1, 3$ $\phi(5) = 4: 1, 2, 3, 4$ $\phi(6) = 2: 1, 5$
$\phi(7) = 6: 1, 2, 3, 4, 5, 6$ $\phi(8) = 4: 1, 3, 5, 7$ $\phi(9) = 6: 1, 2, 4,$
                                                                    $5, 7, 8$

$\phi(10) = 4: 1, 3, 7, 9$ $\phi(11) = 10: 1, 2, \ldots, 10$ $\phi(12) = 4: 1, 5, 7, 11$
$\phi(13) = 12: 1, 2, \ldots, 12$ $\phi(14) = 6: 1, 3, 5, 9, 11, 13$
5. If p is a prime then $\phi(p) = p - 1$. All the numbers $1, 2, \ldots, (p - 1)$ are relatively prime to p.
6. For a power of 2, $\phi(2^n) = 2^{n-1}$, for $n > 0$. This is true since one half the numbers $\leq 2^n$ are odd and relatively prime to 2^n.
7. The results are equal in a, c, f, h, and i. If two numbers are relatively prime then the Euler ϕ number of the product of the numbers can be found from the product of the ϕ numbers for the two numbers.

EXERCISE 8

1. $\{1, 2, 4\},$
     sum $= 7$, deficient
3. $\{1\},$
     sum $= 1$, deficient
5. $\{1, 2, 3, 5, 6, 10, 15\},$
     sum $= 42$, abundant
7. $\{1, 2, 3, 4, 5, 6, 10, 12, 15, 20, 30\},$
     sum $= 108$, abundant
9. $\{1, 2, 4, 5, 10, 20, 25, 50\},$
     sum $= 117$, abundant
11. $\{1, 5, 31\},$
     sum $= 37$, deficient

2. $\{1, 2, 4, 5, 10\},$
     sum $= 22$, abundant
4. $\{1, 2, 3, 4, 6, 8, 12\},$
     sum $= 36$, abundant
6. $\{1, 2, 4, 7, 8, 14, 28\},$
     sum $= 64$, abundant
8. $\{1\},$
     sum $= 1$, deficient
10. $\{1, 2, 4, 37, 74\},$
     sum $= 118$, deficient
12. $\{1, 2, 3, 6, 31, 62, 93\},$
     sum $= 198$, abundant

EXERCISE 9

1. 12, 18, 20, 24, 30, 36, 40, 42, 48, 54, 56, 60, 66, 70, 72, 78, 80, 84, 88, 90, 96, 100, 102, 104, 108, 112, 114, 120, 126, 132, 138, 140, 144, 150, 156, 160, 162, 168, 174, 176, 180, 186, 192, 196, 198, 200, 204, 208, 210, 216. All are even.
2. 945 and 1575 are the first and second odd abundant numbers.

EXERCISE 10

1. a) 36; {1, 2, 3, 4, 6, 9, 12, 18}, sum = 55, abundant
   b) 38; {1, 2, 19}, sum = 22, not abundant
   c) 48; {1, 2, 3, 4, 6, 8, 12, 16, 24}, sum = 76, abundant
   d) 44; {1, 2, 4, 11, 22}, sum = 40, not abundant
   e) 50; {1, 2, 5, 10, 25}, sum = 43, not abundant
   f) 54; {1, 2, 3, 6, 9, 18, 27}, sum = 66, abundant

2. Sometimes. 1 a and b show that the sum may or may not be abundant.

EXERCISE 11

All sums are abundant.

EXERCISE 12

1. a) 32; {1, 2, 4, 8, 16}, sum = 31, deficient
   b) 38; {1, 2, 19}, sum = 22, deficient
   c) 44; {1, 2, 4, 11, 22}, sum = 40, deficient
   d) 59; {1}, sum = 1, deficient
   e) 64; {1, 2, 4, 8, 16, 32}, sum = 63, deficient
   f) 72; {1, 2, 3, 4, 6, 8, 9, 12, 18, 24, 36}, sum = 123, not deficient

2. Sometimes. 1 a and f show that the sum may or may not be deficient.

EXERCISE 13

1. a) 26; {1, 2, 13}, sum = 16, not abundant
   b) 34; {1, 2, 17}, sum = 20, not abundant
   c) 36; {1, 2, 3, 4, 6, 9, 12, 18}, sum = 55, abundant
   d) 56; {1, 2, 4, 7, 8, 14, 28}, sum = 64, abundant
   e) 44; {1, 2, 4, 11, 22}, sum = 40, not abundant
   f) 58; {1, 2, 29}, sum = 32, not abundant

2. Sometimes. 1 a and c show that the sum may or may not be abundant.

EXERCISE 14

1. a) 18; {1, 2, 3, 6, 9}, sum = 21, abundant
   b) 10; {1, 2, 5}, sum = 8, not abundant
   c) 30; {1, 2, 3, 5, 6, 10, 15}, sum = 42, abundant
   d) 8; {1, 2, 4}, sum = 7, not abundant
   e) 16; {1, 2, 4, 8}, sum = 15, not abundant
   f) 44; {1, 2, 4, 11, 22}, sum = 40, not abundant

2. Sometimes. 1 a and b show that the difference may or may not be abundant.

EXERCISE 15

1. a) 27; {1, 3, 9}, sum = 13, deficient

b) 8; {1, 2, 4}, sum = 7, deficient

c) 18; {1, 2, 3, 6, 9}, sum = 21, not deficient

d) 33; {1, 3, 11}, sum = 15, deficient

e) 48; {1, 2, 3, 4, 6, 8, 12, 16, 24}, sum = 76, not deficient

2. Sometimes. 1 a and c show that the difference may or may not be deficient.

EXERCISE 16

1. a) 33; {1, 3, 11}, sum = 15, deficient

b) 30; {1, 2, 3, 5, 6, 10, 15}, sum = 42, abundant

c) 49; {1, 7}, sum = 8, deficient

d) 107; {1}, sum = 1, deficient

e) 140; {1, 2, 4, 5, 7, 10, 14, 20, 28, 35, 70}, sum = 196, abundant

f) 112; {1, 2, 4, 7, 8, 14, 16, 28, 56}, sum = 136, abundant

g) 21; {1, 3, 7}, sum = 11, deficient

h) 40; {1, 2, 4, 5, 8, 10, 20}, sum = 50, abundant

i) 39; {1, 3, 13}, sum = 17, deficient

j) 54; {1, 2, 3, 6, 9, 18, 27}, sum = 66, abundant

k) 104; {1, 2, 4, 8, 13, 26, 52}, sum = 106, abundant

2. Sometimes. 1 a and b show that the difference may or may not be deficient.

EXERCISE 17

1. All products are abundant.

2. Always. In fact, the product of any positive integer and an abundant number is abundant. Let A be an abundant number with proper divisors $d_1, d_2, \ldots, d_k$ such that $d_1 + d_2 + \cdots + d_k > A$. Let N be a positive integer. Then, $N(d_1 + d_2 + \cdots + d_k) > NA$. Since d_i divides A, Nd_i divides NA for all i, so NA is abundant.

3. All products are abundant.

4. Always. See explanation in 2.

5. Infinite. The set of multiples of a number is infinite. Thus, the set of multiples of abundant numbers are abundant and infinite.

6. a, b, d, g, and h are deficient.

7. 6 a and c show the product may or may not be deficient.

EXERCISE 18

1. All multiples are abundant.

2. Always. Take $6m$, $m > 2$. The proper divisors of $6m$ include m, $2m$, $3m$, and 2, whose sum $6m + 2 > 6m$.

3. All multiples are abundant.

4. Always. Take $28m$, $m > 2$. The proper divisors of $28m$ include m, $2m$, $4m$, $7m$, $14m$, and 2 whose sum is $28m + 2 > 28m$.

EXERCISE 19

1. a, b, d, e, g, h, i, j, and k are abundant.
2. Sometimes. For example, 3×4 is abundant, 13×14 is not.
3. All products are abundant and multiples of 6.
4. Always. The product of three consecutive integers is a multiple of the perfect number 6. In every set of three consecutive numbers there is exactly one multiple of 3 and at least one multiple of 2. Thus the product of three such numbers is a multiple of 6 and thus abundant.
5. a) 20×6 b) 840×6 c) 15840×6 d) 123760×6
6. a) Yes, because the product of three consecutive integers is a multiple of 6.
   b) Yes, because all multiples of 6 are abundant.

EXERCISE 20

1. a) 48; $12 + 36, 18 + 30, 24 + 24$
   b) 66; $12 + 54, 18 + 48, 24 + 42, 30 + 36$
   c) 60; $12 + 48, 18 + 42, 20 + 40, 24 + 36, 30 + 30$
   d) 84; $12 + 72, 18 + 66, 24 + 60, 30 + 54, 36 + 48, 42 + 42$
   e) 90; $12 + 78, 18 + 72, 20 + 70, 24 + 66, 30 + 60, 36 + 54, 42 + 48$
   f) 96; $12 + 84, 18 + 78, 24 + 72, 30 + 66, 36 + 60, 40 + 56, 42 + 54,$
      $48 + 48$
   g) 108; $12 + 96, 18 + 90, 20 + 88, 24 + 84, 30 + 78, 36 + 72, 42 + 66,$
      $48 + 60, 54 + 54$
2. a) 36; $12 + 12 + 12$
   b) 48; $12 + 12 + 24, 12 + 18 + 18$
   c) 54; $12 + 12 + 30, 12 + 18 + 24, 18 + 18 + 18$
   d) 60; $12 + 12 + 36, 12 + 18 + 30, 12 + 24 + 24, 18 + 18 + 24,$
      $20 + 20 + 20$
   e) 80; $12 + 12 + 56, 12 + 20 + 48, 18 + 20 + 42, 20 + 20 + 40,$
      $20 + 24 + 36, 20 + 30 + 30$
   f) 72; $12 + 12 + 48, 12 + 18 + 42, 12 + 20 + 40, 12 + 24 + 36,$
      $12 + 30 + 30, 18 + 18 + 36, 18 + 24 + 30, 24 + 24 + 24$
   g) 78; $12 + 12 + 54, 12 + 18 + 48, 12 + 24 + 42, 12 + 30 + 36,$
      $18 + 18 + 42, 18 + 20 + 40, 18 + 24 + 36, 18 + 30 + 30, 24 + 24 + 30$

EXERCISE 21

1.

Power of 2	Proper Divisors	Sum of Proper Divisors	Sum = number -1
$2^4 = 16$	$1, 2, 4, 8$	15	$15 = 2^4 - 1$
$2^5 = 32$	$1, 2, 4, 8, 16$	31	$31 = 2^5 - 1$

Power of 2	Proper Divisors	Sum of Proper Divisors	Sum = number −1
$2^6 = 64$	1, 2, 4, 8, 16, 32	63	$63 = 2^6 - 1$
$2^7 = 128$	1, 2, 4, 8, 16, 32, 64	127	$127 = 2^7 - 1$
$2^8 = 256$	1, 2, 4, 8, 16, 32, 64, 128	255	$255 = 2^8 - 1$
$2^9 = 512$	1, 2, 4, 8, 16, 32, 64, 128, 256	511	$511 = 2^9 - 1$
$2^{10} = 1024$	1, 2, 4, 8, 16, 32, 64, 128, 256, 512	1023	$1023 = 2^{10} - 1$

2. Yes, since the sum of the proper divisors is 1 less than the power of 2.

3. a) 1   b) less than   c) deficient   d) Number of deficient numbers is infinite.

EXERCISE 22

1. All powers of primes are deficient.

2. All powers of primes are deficient.

EXERCISE 23

1. All products are deficient numbers.

2. None. Excepting 2×3, all products of primes are deficient.

EXERCISE 24

1. a) 0, 2, 4, 6, 8   b) 1, 3, 5, 7, 9

2.

Addition				Multiplication		
+	even	odd		×	even	odd
even	even	odd		even	even	even
odd	odd	even		odd	even	odd

3. a) 496   b) 8128

4. a) $2^0(2^1 - 1)$, not even, not perfect   b) $2^1(2^2 - 1)$, even, perfect

c) $2^2(2^3 - 1)$, even, perfect   d) $2^3(2^4 - 1)$, even, not perfect

e) $2^4(2^5 - 1)$, even, perfect   f) $2^5(2^6 - 1)$, even, not perfect

g) $2^6(2^7 - 1)$, even, perfect

EXERCISE 25

1. 360, yes, multiplicity 3   **2.** 465, no   **3.** 1020, no   **4.** 2016, yes, multiplicity 3

EXERCISE 26

1. Yes, +1 type   **2.** No   **3.** Yes, +1 type   **4.** Yes, +1 type   **5.** No

6. No   **7.** No   **8.** Yes, +1 type   **9.** Yes, +1 type

From **EXERCISE 21** it follows that from 2^1 to any power of 2 > 1, we have almost perfect numbers of type +1. So far *no* numbers of type −1 have been found.

EXERCISE 27

All numbers in problems 1−10 are semiperfect.

1. $\{1, 2, 3, 6, 9\}, 9 + 6 + 3$
2. $\{1, 2, 4, 5, 10\}, 10 + 5 + 4 + 1$
3. $\{1, 2, 3, 4, 6, 8, 12\}, 12 + 8 + 4$
4, $\{1, 2, 3, 5, 6, 10, 15\}, 15 + 10 + 5$
5. $\{1, 2, 3, 4, 6, 9, 12, 18\}, 18 + 12 + 6$
6. $\{1, 2, 4, 5, 8, 10, 20\}, 20 + 10 + 8 + 2$
7. $\{1, 2, 3, 6, 7, 14, 21\}, 21 + 14 + 7$
8. $\{1, 2, 3, 4, 6, 8, 12, 16, 24\}, 24 + 16 + 8$
9. $\{1, 2, 4, 7, 8, 14, 28\}, 28 + 14 + 8 + 4 + 2$
10. $\{1, 2, 3, 4, 5, 6, 10, 12, 15, 20, 30\}, 30 + 20 + 10$
11. $\{1, 2, 5, 7, 10, 14, 35\}, 70$ is abundant but not semiperfect.

EXERCISE 28

1. a) $\{1, 2, 3, 6, 9\}, 1 + 2 + 6 + 9, 3 + 6 + 9$
b) $\{1, 2, 4, 5, 10\}, 1 + 4 + 5 + 10$
c) $\{1, 2, 3, 4, 6, 8, 12\}, 1 + 2 + 3 + 4 + 6 + 8,$
   $1 + 2 + 3 + 6 + 12, 4 + 8 + 12, 1 + 3 + 8 + 12,$
   $2 + 4 + 6 + 12$
d) $\{1, 2, 3, 5, 6, 10, 15\}, 1 + 3 + 5 + 6 + 15, 2 + 3 + 10 + 15, 5 + 10 + 15$
e) $\{1, 2, 3, 4, 6, 9, 12, 18\}, 3 + 6 + 9 + 18, 1 + 2 + 3 + 12 + 18,$
   $6 + 12 + 18, 2 + 4 + 12 + 18, 2 + 3 + 4 + 9 + 18,$
   $1 + 2 + 6 + 9 + 18, 2 + 3 + 4 + 6 + 9 + 12$
2. $PD_{72} = \{1, 2, 3, 4, 6, 8, 9, 12, 18, 24, 36\}$

(36, 24, 12)	(36, 18, 9, 6, 3)	(24, 18, 12, 9, 8, 1)
(36, 24, 9, 3)	(36, 18, 9, 6, 2, 1)	(24, 18, 12, 9, 6, 3)
(36, 24, 9, 2, 1)	(36, 18, 9, 4, 3, 2)	(24, 18, 12, 9, 6, 2, 1)
(36, 24, 8, 4)	(36, 18, 8, 6, 4)	(24, 18, 12, 9, 4, 3, 1)
(36, 24, 8, 3, 1)	(36, 18, 8, 6, 3, 1)	(24, 18, 12, 8, 6, 4)
(36, 24, 6, 4, 2)	(36, 18, 8, 4, 3, 2, 1)	(24, 18, 12, 8, 6, 3, 1)
(36, 24, 6, 3, 2, 1)	(36, 12, 9, 8, 6, 1)	(24, 18, 12, 8, 4, 3, 2, 1)
(36, 18, 12, 6)	(36, 12, 9, 8, 4, 3)	(24, 18, 9, 8, 6, 4, 2, 1)
(36, 18, 12, 4, 2)	(36, 12, 9, 8, 4, 2, 1)	(24, 18, 9, 8, 6, 4, 3)
(36, 18, 12, 3, 2, 1)	(36, 12, 9, 6, 4, 3, 2)	
(36, 18, 9, 8, 1)	(36, 12, 8, 6, 4, 3, 2, 1)	

$PD_{96} = \{1, 2, 3, 4, 6, 8, 12, 16, 24, 32, 48\}$

(48, 32, 16)	(48, 16, 12, 8, 6, 3, 2, 1)	(48, 24, 12, 6, 3, 2, 1)
(48, 32, 12, 4)	(48, 24, 16, 8)	(48, 24, 8, 6, 4, 3, 2, 1)
(48, 32, 12, 3, 1)	(48, 24, 16, 6, 2)	(32, 24, 16, 12, 8, 4)
(48, 32, 8, 6, 2)	(48, 24, 16, 4, 3, 1)	(32, 24, 16, 12, 8, 3, 1)

(48, 32, 8, 4, 3, 1)   (48, 24, 12, 8, 4)   (32, 24, 16, 12, 6, 4, 2)
(48, 32, 6, 4, 3, 2, 1) (48, 24, 12, 8, 3, 1)   (32, 24, 16, 12, 6, 3, 2, 1)
(48, 16, 12, 8, 6, 4, 2) (48, 24, 12, 6, 4, 2)   (32, 24, 16, 8, 6, 4, 3, 2,1)

EXERCISE 29

1. 6 is not semiperfect. All the other numbers in this problem are semiperfect.

2. Yes. Proof: Let $m = 6n$ where $n > 1$. The proper divisors of $6n = 2(3)(n)$ include 1, 2, 3, 2(3), n, $2n$, $3n$. But $n + 2n + 3n = 6n = m$.

3. Yes. This follows from problem 2 because all multiples of 6 greater than 6 are semiperfect and the set of multiples of a number is infinite.

4. The 2, 3, 4, 5, ... multiples of the perfect number 28 are semiperfect numbers.
For $n > 1$, the n multiples of a perfect number are semiperfect.

5. The product of 3, 4, 5, or more consecutive numbers are multiples of the perfect number 6. From problem 2, $6n$ for $n > 1$ is a semiperfect number. Thus, the product of 3, 4, 5, or more consecutive numbers will be semiperfect numbers.

6. a) 66, semiperfect   b) 32, not semiperfect   c) 90, semiperfect
d) 44, not semiperfect   e) 82, not semiperfect   f) 120, semiperfect
g) 170, not semiperfect h) 228, semiperfect

7. Sometimes.

8. a) 24, semiperfect   b) 6, not semiperfect   c) 16, not semiperfect
d) 46, not semiperfect   e) 60, semiperfect   f) 84, semiperfect
g) 30, semiperfect   h) 52, not semiperfect

9. Sometimes. See 8 a and b. The difference may or may not be semiperfect.
For further information on weird abundant numbers refer to J. Sándor, D. S. Mitrinović, B. Crstici, eds. (2006). **Handbook of Number Theory I**. Dordrecht: Springer-Verlag. pp. 113–114.

EXERCISE 30

1. a) Not primitive, 18 b) Primitive   c) Primitive   d) Primitive
e) Primitive   f) Not primitive, 48 g) Not primitive, 20   h) Primitive
i) Primitive   j) Not primitive, 30 k) Primitive
l) Not primitive, 20

EXERCISE 31

1. In an amicable pair one number is less than the other, say $a < b$. If b is abundant the sum of the proper divisors would be greater than b and consequently greater than a (rather than equal to a).

2. In an amicable pair one number is less than the other, say $a < b$. If a is *deficient then* the sum of the proper divisors would be less than a and consequently less than b (rather than equal to b).

For more on amicable numbers refer to the work of Herman J.J. te Riele in the journal *Mathematics of Computation.*

EXERCISE 32

Each of the pairs is imperfectly amicable in 1–7.

EXERCISE 33

1. a) $3 \to 1$, 1 link

b) $9 \to 4 \to 3 \to 1$, 3 links

c) $15 \to 9 \to 4 \to 3 \to 1$, 4 links

d) $19 \to 1$, 1 link

e) $20 \to 22 \to 14 \to 10 \to 8 \to 7 \to 1$, 6 links

f) $24 \to 36 \to 55 \to 17 \to 1$, 4 links

g) $25 \to 6$, 1 link

h) $27 \to 13 \to 1$, 2 links

i) $28 \to 28$, 1 link

j) $29 \to 1$, 1 link

k) $30 \to 42 \to 54 \to 66 \to 78 \to 90 \to 144 \to 259 \to 45 \to 33 \to 15 \to$
$9 \to 4 \to 3 \to 1$, 14 links

l) $32 \to 31 \to 1$, 2 links

m) $33 \to 15 \to 9 \to 4 \to 3 \to 1$, 5 links

n) $34 \to 20 \to 22 \to 14 \to 10 \to 8 \to 7 \to 1$, 7 links

o) $38 \to 22 \to 14 \to 10 \to 8 \to 7 \to 1$, 6 links

p) see (k), 13 links

q) see (k), 6 links

r) $46 \to 26 \to 16 \to 15 \to 9 \to 4 \to 3 \to 1$, 7 links

s) see (k), 12 links

t) $60 \to 108 \to 172 \to 136 \to 134 \to 70 \to 74 \to 40 \to 50 \to 43 \to 1$,
10 links

u) $95 \to 25 \to 6$, 2 links

2. One

3. None

For more information on sociable numbers see **Martin Gardners Mathematical Games: The Entire Collection of His** *Scientific American Columns on* **One CD** — in particular see the item *Mathematical Magic Show-*, Washington, D.C.: Mathematical Association of America, 2005, and Malcolm Lines, **A Number for Your Thoughts**, Philadelphia, PA: Adam Hilger, 1988.

EXERCISE 34

1. All of the powers of 2 are practical numbers.

2. Unlimited, because all positive powers of 2 are practical numbers.

3. None of the powers of primes are practical numbers.

4. No.

5. None of the odd composites are practical numbers.

6. No. The number 2 cannot be written as a sum of the distinct divisors of an odd number.

7. a) Practical b) Not practical c) Practical d) Practical e) Not practical
f) Not practical g) Not practical h) Practical i) Practical j) Not practical
k) Practical l) Not practical m) Practical n) Practical o) Not practical

8. Answers will vary.

a) 12, 36, 40, 42, 54, 56, 60, 66, 72, 78

b) 70, 102, 114, 138, 174, 186, 222, 246, 258, 282.

For more information on practical numbers, see A. K. Srinivasan, **Current Science**, Vol. 17 (June, 1948): pp. 179–180.

EXERCISE 35

1. $\{1, 2, 4, 7, 14, 28\}$, sum $= 56$

2.

$1 = 1$	$2 = 2$	$3 = 1 + 2$
$4 = 4$	$5 = 1 + 4$	$6 = 2 + 4$
$7 = 7$	$8 = 1 + 7$	$9 = 2 + 7$
$10 = 1 + 2 + 7$	$11 = 4 + 7$	$12 = 1 + 4 + 7$
$13 = 2 + 4 + 7$	$14 = 1 + 2 + 4 + 7$	$15 = 1 + 14$
$16 = 2 + 14$	$17 = 1 + 2 + 14$	$18 = 4 + 14$
$19 = 1 + 4 + 14$	$20 = 2 + 4 + 14$	$21 = 1 + 2 + 4 + 14$
$22 = 1 + 7 + 14$	$23 = 2 + 7 + 14$	$24 = 1 + 2 + 7 + 14$
$25 = 4 + 7 + 14$	$26 = 1 + 4 + 7 + 14$	$27 = 2 + 4 + 7 + 14$
$28 = 1 + 2 + 4 + 7 + 14$		

3. 28 is practical as shown in problem 2 solution.

4. It is possible — add 28 to each of the sums in problem 2 solution.

EXERCISE 36

Answers will vary in all cases.

1. (1, 2, 4, 6, 12), (1, 2, 4, 7, 11), (1, 2, 3, 7, 12)

2. (1, 2, 4, 8, 12), (1, 2, 4, 6, 14), (1, 2, 3, 7, 14), (1, 2, 4, 7, 13)

3. (1, 1, 2, 5, 10, 20), (1, 1, 3, 4, 10, 20), (1, 1, 3, 5, 9, 20), (1, 1, 3, 5, 10, 19), (1, 1, 3, 5, 11, 18)

4. (1, 1, 2, 5, 10, 20, 40), (1, 1, 3, 4, 10, 20, 40), (1, 1, 3, 5, 9, 20, 40), (1, 1, 3, 5, 10, 19, 40), (1, 1, 3, 5, 10, 20, 39)

5. (1, 1, 2, 5, 10, 20, 40, 80), (1, 1, 3, 4, 10, 20, 40, 80), (1, 1, 3, 5, 9, 20, 40, 80), (1, 1, 3, 5, 10, 19, 40, 80), (1, 1, 3, 5, 10, 20, 39, 80)

EXERCISE 37

1. Exploration

2. a) $60°$ b) $90°$ c) $108°$ d) $120°$

3. Exploration

EXERCISE 38

1. a) $1 + 2 + 3 + \cdots + 6 = 21$ b) $1 + 2 + 3 + \cdots + 8 = 36$

c) $1 + 2 + 3 + \cdots + 10 = 55$ d) $1 + 2 + 3 + \cdots + n = \dfrac{n(n+1)}{2}$

2. a) 35   b) 67   c) 100   d) n

EXERCISE 39

1. Use formula for triangular numbers with $n = 1, 2, \ldots, 30$. The results are 1, 3, 6, 10, 15, 21, 28, 36, 45, 55, 66, 78, 91, 105, 120, 136, 153, 171, 190, 210, 231, 253, 276, 300, 325, 351, 378, 406, 435, 465

2. Infinite, because there are an infinite number of integers n to apply in the formula.

EXERCISE 40

1. a) Sometimes   b) Sometimes   c) Sometimes

2. a) $10 + 1$ b) $10 + 3 + 1$ c) $28 + 6$ d) $36 + 1$ e) $36 + 3 + 1$

f) $28 + 15$ g) $45 + 6$ h) $55 + 10 + 3$ i) $66 + 6 + 3$ j) $78 + 3 + 1$

k) $91 + 3$ l) $55 + 45$

3. Three

4. a) $15 + 15 + 6$ b) $45 + 10$ c) $36 + 36 + 6$ d) $55 + 36$

e) $105 + 15$ f) $105 + 45 + 3$ g) $153 + 36 + 1$ h) $210 + 21$

i) $276 + 21 + 3$

5. Three

EXERCISE 41

1. $1 + 3 + 3 + 21$ **2.** $1 + 1 + 15 + 28$ **3.** $1 + 1 + 28 + 36$

4. $1 + 3 + 10 + 91$ **5.** $3 + 6 + 36 + 91$ **6.** $1 + 6 + 28 + 136$

7. $1 + 1 + 55 + 153$ **8.** $1 + 1 + 21 + 253$ **9.** $1 + 3 + 21 + 300$

EXERCISE 42

1. a) $10^2 = 100 = 1^3 + 2^3 + 3^3 + 4^3 = (1 + 2 + 3 + 4)^2$

b) $15^2 = 225 = 1^3 + 2^3 + \cdots + 5^3 = (1 + 2 + \cdots + 5)^2$

c) $21^2 = 441 = 1^3 + 2^3 + \cdots + 6^3 = (1 + 2 + \cdots + 6)^2$

d) $28^2 = 784 = 1^3 + 2^3 + \cdots + 7^3 = (1 + 2 + \cdots + 7)^2$

e) $36^2 = 1296 = 1^3 + 2^3 + \cdots + 8^3 = (1 + 2 + \cdots + 8)^2$

f) $45^2 = 2025 = 1^3 + 2^3 + \cdots + 9^3 = (1 + 2 + \cdots + 9)^2$

g) $55^2 = 3025 = 1^3 + 2^3 + \cdots + 10^3 = (1 + 2 + \cdots + 10)^2$

2. For certain values of n, T_n, the nth triangular number, is also a perfect number. Take the triangular number $T_n = \dfrac{n(n+1)}{2}$ and set $n = (2^p - 1)$. Then $T_{2^p-1} = \dfrac{(2^p - 1)(2^p - 1 + 1)}{2} = \dfrac{2^p(2^p - 1)}{2}$ so that $T_{2^p-1} = 2^{p-1}(2^p - 1)$. Thus, for the values $n = 2^p - 1$, where p is prime and $(2^p - 1)$ is a Mersenne prime, the triangular number T_{2^p-1} is a perfect number. That is, $T_{2^p-1} = 2^{p-1}(2^p - 1)$ is Euclid's formula for even perfect numbers. Thus, every even perfect number is also a triangular number.

EXERCISE 43

1. Row 5: 1, 5, 10, 10, 5, 1   Row 6: 1, 6, 15, 20, 15, 6, 1
Row 7: 1, 7, 21, 35, 35, 21, 7, 1   Row 8: 1, 8, 28, 56, 70, 56, 28, 8, 1
2. There are two similar views of diagonals: left and right. The first diagonal is all 1s, the second diagonal displays the natural numbers, the third diagonal displays the triangular numbers, and so on.
3.

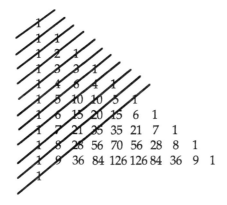

The sums along the diagonals examined are 1, 1, 2, 3, 5, 8, 13, 21, 34, 55, 89, the Fibonacci numbers.

EXERCISE 44

Answers will vary
1. (1, 1, 1), (5, 5, 5), (6, 6, 6), (7, 7, 7)
2. (2, 2, 3), (3, 3, 4), (4, 4, 3), (5, 5, 9)
3. (3, 5, 6), (4, 7, 10), (5, 6, 10), (4, 6, 9)

EXERCISE 45

1.

1, 2, 3	3	4	0	7
1, ... , 4	4	8	1	13
1, ... , 5	5	14	3	22
1, ... , 6	6	21	7	34
1, ... , 7	7	30	13	50
1, ... , 8	8	40	22	70
1, ... , 9	9	52	34	95
1, ... , 10	10	65	50	125

2. Given the lengths corresponding to $1, 2, 3, 4, 5, \ldots, n$, then

a) There are n equilateral triangles

b) For the number of isosceles triangles there are two formulas.

      i) If n is even, set $n = 2m$ and there are $3m^2 - 2m$ isosceles triangles. For example, take $1, 2, 3, 4$, then $n = 4$ and is even. Write $n = 4 = 2(2)$ so that $m = 2$. Then from $3m^2 - 2m$ there are $3(2^2) - 2(2) = 12 - 4 = 8$ isosceles triangles.

      ii) If n is odd, set $n = 2m + 1$ and there are $3m^2 + m$ isosceles triangles. For example, take $1, 2, 3, 4, 5$ then $n = 5$ and is odd. Write $n = 5 = 2(2) + 1$ so that $m = 2$. From $3m^2 + m$ there are $3(2^2) + 2 = 14$ isosceles triangles.

c) For the number of scalene triangles there are two formulas.

      i) If n is even, set $n = 2m$, where $m > 1$ and there are $\dfrac{m(4m - 5)(m - 1)}{6}$ scalene triangles. For example, take $1, 2, 3, 4$ then $n = 4$ and is even. Write $n = 4 = 2(2)$ and $m = 2$. Using the formula $\dfrac{2(4(2) - 5)(2 - 1)}{6} = 1$ scalene triangle.

      ii) If n is odd, set $n = 2m + 1$, and there are $\dfrac{m(4m + 1)(m - 1)}{6}$ scalene triangles. For example, take $1, 2, 3, 4, 5$ then $n = 5$ and is odd. Write $n = 5 = 2(2) + 1$ and $m = 2$. From the formula, $\dfrac{2(4(2) + 1)(2 - 1)}{6} = 3$ scalene triangles.

d) For the total number of equilateral, isosceles and scalene triangles there are two formulas.

      i) If n is even, set $n = 2m$ and there are $\dfrac{m(4m + 5)(m + 1)}{6}$ triangles in all. For example, take $1, 2, 3, 4$ then $n = 4 = 2(2)$ and $m = 2$. From the formula there are $\dfrac{2(4(2) + 5)(2 + 1)}{6} = 13$ triangles in all.

      ii) If n is odd, set $n = 2m + 1$ and there are $\dfrac{(m + 2)(4m + 3)(m + 1)}{6}$ triangles in all. For example, take $1, 2, 3, 4, 5$ then $n = 5 = 2(2) + 1$ and $m = 2$. From the formula there are $\dfrac{(2 + 2)(4(2) + 3)(2 + 1)}{6} = 22$ triangles in all.

The formulas above were developed using the method of finite differences.

3.

$1, 2, 3, \ldots, 9$	$T_2 + T_4 + T_6 = 3 + 10 + 21 = 34$
$1, 2, 3, \ldots, 10$	$T_1 + T_3 + T_5 + T_7 = 50$
$1, 2, 3, \ldots, 11$	$T_2 + T_4 + T_6 + T_8 = 70$
$1, 2, 3, \ldots, 12$	$T_1 + T_3 + T_5 + T_7 + T_9 = 95$
$1, 2, 3, \ldots, 13$	$T_2 + T_4 + T_6 + T_8 + T_{10} = 125$
$1, 2, 3, \ldots, 14$	$T_1 + T_3 + T_5 + T_7 + T_9 + T_{11} = 161$

4.

$10, 11, 12$	3	6	1	10
$5, 6, 7, 8$	4	12	4	20
$8, 9, \ldots, 12$	5	20	10	35
$15, 16, \ldots, 20$	6	30	20	56
$21, 22, \ldots, 27$	7	42	35	84
$45, 46, \ldots, 52$	8	56	56	120

EXERCISE 46

1. a) $2 + 4 + 6 + 8 + 10 = 5 \times 6 = 30$ b) $2 + 4 + \cdots + 12 = 6 \times 7 = 42$
c) $2 + 4 + \cdots + 14 = 7 \times 8 = 56$ d) $2 + 4 + \cdots + 16 = 8 \times 9 = 72$
e) $2 + 4 + \cdots + 18 = 9 \times 10 = 90$ f) $2 + 4 + \cdots + 20 = 10 \times 11 = 110$
g) $2 + 4 + \cdots + 24 = 12 \times 13 = 156$ h) $2 + 4 + \cdots + 30 = 15 \times 16 = 240$
i) $2 + 4 + \cdots + 36 = 18 \times 19 = 342$ j) $2 + 4 + \cdots + 40 = 20 \times 21 = 420$

2. a) $2 \times 15 = 30$ b) $2 \times 21 = 42$ c) $2 \times 28 = 56$ d) $2 \times 36 = 72$
e) $2 \times 45 = 90$ f) $2 \times 55 = 110$ g) $2 \times 66 = 132$ h) $2 \times 78 = 156$
i) $2 \times 91 = 182$ j) $2 \times 105 = 210$ k) $2 \times 120 = 240$ l) $2 \times 136 = 272$

3. Answers will vary. $(1 \times 2, 3 \times 6), (3 \times 4, 6 \times 8), (4 \times 5, 8 \times 10), (8 \times 10, 16 \times 20),$
$(3 \times 5, 6 \times 10)$
For more information on rectangular numbers, see Euclid, D. Densmore (Editor) Sir T. L. Heath (Translator), **Euclid's Elements**, Santa Fe, NM: Green Lion Press, 2002. See Book 8, Prop. 18.

EXERCISE 47

1. a) 196 b) 324 c) 441 d) 625
2. Unlimited, because the number of integers is unlimited.
3. a) $1 + 3$, 2 terms b) $1 + 3 + 5$, 3 terms
c) $1 + 3 + \cdots + 11$, 6 terms d) $1 + 3 + \cdots + 19$, 10 terms
e) $1 + 3 + \cdots + 23$, 12 terms f) $1 + 3 + \cdots + 39$, 20 terms

4. a) 7 terms, sum $= 49$, 7th b) 8 terms, sum $= 64$, 8th
c) 9 terms, sum $= 81$, 9th d) 12 terms, sum $= 144$, 12th
e) 13 terms, sum $= 169$, 13th f) 16 terms, sum $= 256$, 16th

5. a) $T_4 + T_5 = 10 + 15 = 25$ b) $T_5 + T_6 = 15 + 21 = 36$
   c) $T_6 + T_7 = 21 + 28 = 49$ d) $T_7 + T_8 = 28 + 36 = 64$
   e) $T_8 + T_9 = 36 + 45 = 81$ f) $T_9 + T_{10} = 45 + 55 = 100$
   g) $T_{14} + T_{15} = 105 + 120 = 225$ h) $T_{n-1} + T_n = n^2$

EXERCISE 48

1. $6 = 1^2 + 1^2 + 2^2$, $7 = 1^2 + 1^2 + 1^2 + 2^2$, $8 = 2^2 + 2^2$,
    $9 = 3^2$, $10 = 1^2 + 3^2$, $11 = 1^2 + 1^2 + 3^2$,
    $12 = 1^2 + 1^2 + 1^2 + 3^2$, $13 = 2^2 + 3^2$, $14 = 1^2 + 2^2 + 3^2$,
    $15 = 1^2 + 1^2 + 2^2 + 3^2$, $16 = 4^2$, $17 = 1^2 + 4^2$,
    $18 = 3^2 + 3^2$, $19 = 1^2 + 3^2 + 3^2$

2. a) four   b) four

Read about Bachet's (1591–1639) conjecture that any number is either a square or a sum of two, three, or four squares. Joseph Lagrange (1736–1813) in 1770 was the first to prove that every integer could be written as the sum of at most four squares. Also see L. E. Dickson, **History of the Theory of Numbers**, New York: Chelsea Publishing Co., Vol. 2 (1952), pp. 275–303 and J. D. Dixon. "Another Proof of Lagrange's 4-Square Theorem." *American Math Monthly*, Vol. 71 (1964), p. 286.

3. a) $4^2 + 5^2 = 41 = (10)4 + 1$ b) $5^2 + 6^2 = 61 = (15)4 + 1$
   c) $6^2 + 7^2 = 85 = (21)4 + 1$ d) $7^2 + 8^2 = 113 = (28)4 + 1$
   e) $8^2 + 9^2 = 145 = (36)4 + 1$ f) $9^2 + 10^2 = 181 = (45)4 + 1$
   g) $10^2 + 11^2 = 221 = (55)4 + 1$ h) $14^2 + 15^2 = 421 = (105)4 + 1$
   i) $19^2 + 20^2 = 761 = (190)4 + 1$

4. The sum of two consecutive squares $n^2 + (n + 1)^2$ is equal to 4 times the nth triangular number plus 1, that is, $n^2 + (n + 1)^2 = \dfrac{n(n + 1)}{2}(4) + 1$ where $\dfrac{n(n + 1)}{2}$ is the nth triangular number.

EXERCISE 49

1. a, b, c, d are not square pair numbers.

e)	A:	1	2	3	4	5	6	7	8		
	B:	8	7	6	5	4	3	2	1		
f)	A:	1	2	3	4	5	6	7	8	9	
	B:	8	2	6	5	4	3	9	1	7	
g)	A:	1	2	3	4	5	6	7	8	9	10
	B:	3	2	1	5	4	10	9	8	7	6

2. 3, 5, 8, 9, 10

3. 11 is not a square pair number but all numbers greater than 11 are square pair numbers. One possible strategy to use in matching the numbers 1 through n with themselves is to list under each number in the array 1 through n all those numbers which with that number sum to a perfect square. If for some number there is only a single choice, eliminate that choice from all other positions. If a number appears just once, eliminate any other choices from that position. If there are still multiple choices after doing this make a selection and continue the elimination until there is a one to one match.

Example:

```
A:  1   2   3   4   5   6   7   8   9   10  11  12
B:  3   2   1   8   4   8   2   1   7   6   5   4
            8   1   8   12  11  10  9   8
```

Start by crossing out 4 under 5, then cross out 5 under 4, 6 under 3, 3 under 6, 8 under 1, 1 under 8, 7 under 2, 2 under 7. Another strategy is to begin by finding a match for the largest number and working backwards. Answers are not necessarily unique. Some answers are found readily such as 1 through 24 can be matched with 24 through 1.

EXERCISE 50

1. b) and d)   **2.** Both   **3.** Yes   **4.** Answers will vary.

For more information on bigrade pairs of triples, see P. Pisa. "Note 385, A Remarkable Bigrade." *Scripta Mathematica*, Vol. 20 (1954), p. 213.

EXERCISE 51

1–10. All the triples of numbers are Pythagorean triples.
11. Develop criteria to evaluate reports.

EXERCISE 52

1. All the triples in Exercise 51 are primitive Pythagorean triples.
2. (16, 30, 34): $16^2 + 30^2 = 1156 = 34^2$; (24, 45, 51): $24^2 + 45^2 = 2601 = 51^2$
3. (50, 120, 130): $50^2 + 120^2 = 16900 = 130^2$; (100, 240, 260): $100^2 + 240^2 = 67600 = 260^2$
4. Let (a, b, c) be a Pythagorean triple. The n multiple, (na, nb, nc) is a Pythagorean triple because $(na)^2 + (nb)^2 = n^2a^2 + n^2b^2 = n^2(a^2 + b^2) = n^2(c^2) = (nc)^2$.
5. a) Yes   b) No   c) No   d) Yes   e) Yes
6. Answers will vary. $(84, 13, 85)$, $(65, 72, 97)$, $(48, 55, 73)$, $(17, 144, 145)$, $(15, 112, 113)$
7. $(119, 120, 169)$. The rest of this problem is an open challenge.

EXERCISE 53

1. a) 96; $100 + 96 = 14^2$, $100 - 96 = 2^2$
   b) 240; $289 + 240 = 23^2$, $289 - 240 = 7^2$
   c) 384; $400 + 384 = 28^2$, $400 - 384 = 4^2$
   d) 336; $625 + 336 = 31^2$, $625 - 336 = 17^2$
   e) 840; $841 + 840 = 41^2$, $841 - 840 = 1^2$
   f) 480; $676 + 480 = 34^2$, $676 - 480 = 14^2$
2. a) Yes, $n = 2 \cdot 9 \cdot 12 = 216$
   b) Yes, $n = 2 \cdot 18 \cdot 24 = 864$

EXERCISE 54

1. 1, 7, 10, 13, 19, 23, 28, 31, 32, 44, 49, 68, 70, 79, 82, 86, 91, 94, 97

EXERCISE 55

1. Answers will vary.
   a) $100 + 130 = 230$ is happy.   b) $70 + 10 = 80$ is not happy.
   c) $230 - 100 = 130$ is happy.   d) $13 - 10 = 3$ is not happy.
   e) $10 \times 10 = 100$ is happy.   f) $13 \times 13 = 169$ is not happy.
2. Answers will vary.

EXERCISE 56

1. Brian, Hope, Taj, Gina, Huan   **2.** Exploration
3. July, September, November   **4.** Wednesday, Saturday
5. 2008, 2030, 2111, 2190

EXERCISE 57

1. a) 20   b) 37   c) 16   d) 4   e) 58   f) 58   g) 89   h) 89   i) 145

2. With each cycle number on the left, the numbers from 2 through 99 that come into the cycle at that point are indicated. Numbers not in the table are happy.

4	2, 11, 78, 87
16	15, 26, 40, 51, 62, 69, 88, 96
37	3, 9, 18, 30, 33, 39, 47, 56, 57, 59, 61, 65, 74, 75, 81, 90, 93, 95
58	38, 73, 83
89	5, 6, 8, 12, 14, 17, 21, 22, 25, 27, 29, 34, 35, 36, 41, 43, 45, 46, 48, 50, 52, 53, 54, 55, 60, 63, 64, 66, 67, 71, 72, 76, 80, 84, 85, 92, 99
145	77, 98
42	
20	24

EXERCISE 58

1. The middle digit in 121, and 12321 is the number of 1s in 11, and in 111.

2. a) 1234321   b) 123454321   c) 12345654321

    d) 1234567654321   e) 123456787654321   f) 12345678987654321

3. Yes. The pattern breaks down at $1,111,111,111^2$ since carrying is necessary in determining the product.

EXERCISE 59

1. a) $5 + 1$ b) $6 + 1$ c) $7 + 1$ d) $7 + 2$ e) $7 + 3$

   f) $6 + 5$ g) $17 + 15$ h) $22 + 21$ i) $31 + 23$ j) $34 + 31$

2. a) $19 + 2 = 15 + 6 = 14 + 7 = 11 + 10$

   b) $21 + 1 = 19 + 3 = 17 + 5 = 15 + 7 = 11 + 11$

   c) $22 + 1 = 21 + 2 = 17 + 6 = 13 + 10$

   d) $23 + 1 = 22 + 2 = 21 + 3 = 19 + 5 = 17 + 7 = 14 + 10 = 13 + 11$

   e) $23 + 2 = 22 + 3 = 19 + 6 = 15 + 10 = 14 + 11$

3. a) $1^2(10)$ b) $1^2(11)$ c) $2^2(3)$ d) $1^2(13)$ e) $1^2(14)$

   f) $1^2(42)$ g) $2^2(15)$ h) $5^2(7)$ i) $1^2(210)$ j) $7^2(6)$

EXERCISE 60

1. 5th: $1 + 3 + 6 + 9 + 12 = 31$ 6th: $1 + 3 + \cdots + 15 = 46$

   7th: $1 + 3 + \cdots + 18 = 64$ 8th: $1 + 3 + \cdots + 21 = 85$

   9th: $1 + 3 + \cdots + 24 = 109$ 10th: $1 + 3 + \cdots + 27 = 136$

2.

4th: 10	3rd: 9	4th: $10 + 9 = 19$
5th: 15	4th: 16	5th: $15 + 16 = 31$
6th: 21	5th: 25	6th: $21 + 25 = 46$
7th: 28	6th: 36	7th: $28 + 36 = 64$
8th: 36	7th: 49	8th: $36 + 49 = 85$
9th: 45	8th: 64	9th: $45 + 64 = 109$

3. Unlimited, because the number of triangular and square numbers is unlimited and each tetragonal number is the sum of a triangular and square number.

4. 1 4 0 9 1 6 4 5 9 6 6 9 5 4 6 1 9 0 4 1 — a palindrome of 20 digits which then repeats. The digits 2, 3, 7, 8 do not appear in the palindrome.

5. The nth tetragonal number is $1 + \dfrac{3n(n - 1)}{2}$ or $\dfrac{n(n + 1)}{2} + (n - 1)^2$.

EXERCISE 61

1. a) $1 + 1 + 49$ b) $9 + 25 + 36$ c) $1 + 1 + 9 + 81$

   d) $36 + 81$ e) $1 + 144$ f) $16 + 16 + 144$

   g) $16 + 25 + 169$ h) $1 + 1 + 49 + 196$ i) $1 + 9 + 81 + 196$

2. $92 = 9^2 + 3^2 + 1^2 + 1^2$

3. a) $6 + 45$ b) $15 + 55$ c) $1 + 91$ d) $36 + 36 + 45$
   e) $45 + 45 + 55$ f) $55 + 55 + 66$ g) 210 h) $78 + 78 + 91$
   i) $91 + 91 + 105$

4. 22 is pentagonal and is the sum of the two triangular numbers 1 and 21.

5. a) $6 + 6 + 10$ b) $10 + 10 + 15$ c) $15 + 15 + 21$ d) $21 + 21 + 28$
   e) $28 + 28 + 36$ f) $36 + 36 + 45$ g) $45 + 45 + 55$
   h) $55 + 55 + 66$ i) $66 + 66 + 78$ j) $78 + 78 + 91$

6. The nth pentagonal number is the sum of the nth triangular number plus twice the $(n - 1)$st triangular number.

EXERCISE 62

1. 5th: $1 + 5 + 9 + \cdots + 17 = 45$ 6th: $1 + 5 + 9 + \cdots + 21 = 66$
   7th: $1 + 5 + 9 + \cdots + 25 = 91$ 8th: $1 + 5 + 9 + \cdots + 29 = 120$
   9th: $1 + 5 + 9 + \cdots + 33 = 153$ 10th: $1 + 5 + 9 + \cdots + 37 = 190$
   11th: $1 + 5 + 9 + \cdots + 41 = 231$ 12th: $1 + 5 + 9 + \cdots + 45 = 276$
   13th: $1 + 5 + 9 + \cdots + 49 = 325$ 14th: $1 + 5 + 9 + \cdots + 53 = 378$
   15th: $1 + 5 + 9 + \cdots + 57 = 435$ 16th: $1 + 5 + 9 + \cdots + 61 = 496$

2.

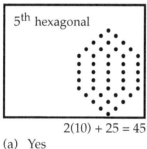

5th hexagonal

$2(10) + 25 = 45$

(a) Yes

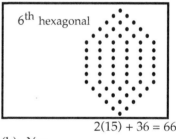

6th hexagonal

$2(15) + 36 = 66$

(b) Yes

3.

5th: 25	4th: 10	5th: 45
6th: 36	5th: 15	6th: 66
7th: 49	6th: 21	7th: 91
8th: 64	7th: 28	8th: 120
9th: 81	8th: 36	9th: 153
10th: 100	9th: 45	10th: 190
11th: 121	10th: 55	11th: 231
12th: 144	11th: 66	12th: 276
13th: 169	12th: 78	13th: 325
14th: 196	13th: 91	14th: 378
15th: 225	14th: 105	15th: 435
16th: 256	15th: 120	16th: 496

4.

4th: 10	5th: 35	5th: 45
5th: 15	6th: 51	6th: 66
6th: 21	7th: 70	7th: 91
7th: 28	8th: 92	8th: 120
8th: 36	9th: 117	9th: 153
9th: 45	10th: 145	10th: 190
10th: 55	11th: 176	11th: 231
11th: 66	12th: 210	12th: 276
12th: 78	13th: 247	13th: 325
13th: 91	14th: 287	14th: 378
14th: 105	15th: 330	15th: 435
15th: 120	16th: 376	16th: 496

5.

4th: 7	4th: 4	4th: 28
5th: 9	5th: 5	5th: 45
6th: 11	6th: 6	6th: 66
7th: 13	7th: 7	7th: 91
8th: 15	8th: 8	8th: 120
9th: 17	9th: 9	9th: 153
10th: 19	10th: 10	10th: 190
11th: 21	11th: 11	11th: 231
12th: 23	12th: 12	12th: 276

6. The circled numbers are hexagonal numbers. Moreover, the circled numbers are in the odd positions $1, 3, 5, 7, 9, \ldots$. It appears that in the formula for triangular numbers, the odd values of n (call them $2k - 1$) yield hexagonal numbers. Substituting $(2k - 1)$ for n in $\dfrac{n(n + 1)}{2}$ gives $\dfrac{(2k - 1)(2k - 1 + 1)}{2} = k(2k - 1)$ which is the formula for hexagonal numbers.

7. Unlimited, because the number of triangular numbers in odd positions is unlimited.

8. 1 6 5 8 5 6 1 0 3 0 — these 10 digits now repeat. Digits that do not appear are 2, 4, 7, and 9.

9. The nth hexagonal number is $n(2n - 1)$.

EXERCISE 63

1. a) $1 + 16 + 49$ b) $1 + 9 + 81$ c) $4 + 16 + 100$
  d) $9 + 144$ e) $9 + 81 + 100$ f) $1 + 9 + 100 + 121$
  g) $4 + 16 + 256$ h) $100 + 225$ i) $1 + 16 + 361$
  j) $25 + 49 + 361$ k) $4 + 4 + 4 + 484$

2. $15 = 3^2 + 2^2 + 1^2 + 1^2$

3. a) $21 + 45$ b) $36 + 55$ c) $15 + 105$ d) $3 + 45 + 105$
    e) $1 + 36 + 153$ f) $21 + 210$ g) $45 + 231$ h) $3 + 91 + 231$
    i) $78 + 300$ j) $1 + 28 + 406$ k) $171 + 325$

4.

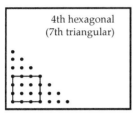 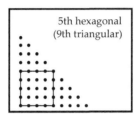 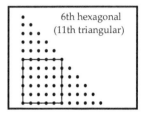

Every hexagonal number is the sum of a square number and twice a triangular number. Hexagonal numbers are also triangular numbers T_n where n is odd.

5. Take the hexagonal number formula, $H_n = n(2n - 1)$ and set $n = 2^{p-1}$.
    Then $H_{2^{p-1}} = 2^{p-1}(2(2^{p-1}) - 1) = 2^{p-1}(2^{p-1+1} - 1)$ so that $H_{2^{p-1}} = 2^{p-1}(2^p - 1)$.

Thus, for the values $n = 2^{p-1}$, if $(2^p - 1)$ is a Mersenne prime, then the hexagonal number $H_{2^{p-1}}$ is a perfect number.

$H_{2^{p-1}} = 2^{p-1}(2^p - 1)$ is Euclid's formula for even perfect numbers.
Thus, every even perfect number is also a hexagonal number.

EXERCISE 64

1, 2.

# of Sides	First Five Terms	Recursive Formula	Closed Formula
3	1 3 6 10 15	$T_n = T_{n-1} + n$	$T_n = \dfrac{n(n+1)}{2}$
4	2 6 12 20 30	$R_n = R_{n-1} + 2n$	$R_n = n(n+1)$
4	1 4 9 16 25	$S_n = S_{n-1} + 2n - 1$	$S_n = n^2$
5	1 5 12 22 35	$P_n = P_{n-1} + 3n - 2$	$P_n = \dfrac{3n^2 - n}{2}$
6	1 6 15 28 45	$H_n = H_{n-1} + 4n - 3$	$H_n = \dfrac{4n^2 - 2n}{2}$
7	1 7 18 34 55	$HP_n = H_{n-1} + 5n - 4$	$HP_n = \dfrac{5n^2 - 3n}{2}$

# of Sides	First Five Terms	Recursive Formula	Closed Formula
8	1 8 21 40 65	$O_n = O_{n-1} + 6n - 5$	$O_n = \dfrac{6n^2 - 4n}{2}$
9	1 9 24 46 75	$N_n = N_{n-1} + 7n - 6$	$N_n = \dfrac{7n^2 - 5n}{2}$
10	1 10 27 52 85	$D_n = D_{n-1} + 8n - 7$	$D_n = \dfrac{8n^2 - 6n}{2}$
11	1 11 30 58 95	$U_n = U_{n-1} + 9n - 8$	$U_n = \dfrac{9n^2 - 7n}{2}$
12	1 12 33 64 105	$DO_n = D_{n-1} + 10n - 9$	$DO_n = \dfrac{10n^2 - 8n}{2}$
13	1 13 36 70 115	$TR_n = T_{n-1} + 11n - 10$	$TR_n = \dfrac{11n^2 - 9n}{2}$
14	1 14 39 76 125	$TE_n = T_{n-1} + 12n - 11$	$TE_n = \dfrac{12n^2 - 10n}{2}$
15	1 15 42 82 135	$PD_n = P_{n-1} + 13n - 12$	$PD_n = \dfrac{13n^2 - 11n}{2}$
16	1 16 45 88 145	$HX_n = H_{n-1} + 14n - 13$	$HX_n = \dfrac{14n^2 - 12n}{2}$
17	1 17 48 94 155	$HD_n = H_{n-1} + 15n - 14$	$HD_n = \dfrac{15n^2 - 13n}{2}$
18	1 18 51 100 165	$OC_n = O_{n-1} + 16n - 15$	$OC_n = \dfrac{16n^2 - 14n}{2}$
19	1 19 54 106 175	$ND_n = N_{n-1} + 17n - 16$	$ND_n = \dfrac{17n^2 - 15n}{2}$
20	1 20 57 112 185	$I_n = I_{n-1} + 18n - 17$	$I_n = \dfrac{18n^2 - 16n}{2}$

EXERCISE 65

1.

	Numbers				Remainders		
Tri.	Square	Pent.	Hex.	Tri.	Square	Pent.	Hex.
4th: 10	16	22	28	1	0	2	4
5th: 15	25	35	45	0	1	0	3

	Numbers				Remainders		
Tri.	Square	Pent.	Hex.	Tri.	Square	Pent.	Hex.
6th: 21	36	51	66	0	0	1	0
7th: 28	49	70	91	1	1	0	1
8th: 36	64	92	120	0	0	2	0
9th: 45	81	117	153	0	1	2	3
10th: 55	100	145	190	1	0	0	4
11th: 66	121	176	231	0	1	1	3
12th: 78	144	210	276	0	0	0	0

2.
a) 1 0 0 1 0 0 ... ; three   b) 1 0 1 0 1 0 ... ; two
c) 1 0 2 2 0 1 0 2 2 0 ... ; five   d) 1 0 3 4 3 0 1 0 3 4 3 0 ... ; six

EXERCISE 66

1. a) 1, 4, 7, 10, 13, 16, 19, 22, 25, 28, ...
b) 1, 7, 13, 19, 25, 31, 37, 43, 49, 55, ...
c) 1, 11, 21, 31, 41, 51, 61, 71, 81, 91, ...
d) 5, 9, 13, 17, 21, 25, 29, 33, 37, 41, ...
e) 4, 9, 14, 19, 24, 29, 34, 39, 44, 49, ...
2. a) 1, 4, 7, 10, 13, 16   b) 1, 5, 9, 13, 17, 21

EXERCISE 67

1.

5th	$20 + 15 = 35$
6th	$35 + 21 = 56$
7th	$56 + 28 = 84$
8th	$84 + 36 = 120$
9th	$120 + 45 = 165$
10th	$165 + 55 = 220$
11th	$220 + 66 = 286$
12th	$286 + 78 = 364$

The second column is an abbreviated computation. Thus, the 3rd tetrahedral number is the sum of the first three triangular numbers $1 + 3 + 6 = 10$ and instead of writing $1 + 3 + 6 = 10$ for the 4th tetrahedral number, take the sum of the first three triangular numbers 10 and add to this the 4th triangular number 10 to get $10 + 10 = 20$, and so on.

2. a) 35   b) 56   c) 84   d) 120   e) 220

f) 680   g) 1140   h) 1540   i) 2925   j) 22,100

3. The formula for tetrahedral numbers comes from the quotient rule in problem 2, namely $\dfrac{n(n+1)(n+2)}{6}$. It can also be derived using the data from the table and applying the method of finite differences.

EXERCISE 68

1.
6th $55 + 36 = 91$
7th $91 + 49 = 140$
8th $140 + 64 = 204$
9th $204 + 81 = 285$
10th $285 + 100 = 385$

2. See the figure preceding Exercise 68. The 5th square pyramidal number will have in the 5th layer 15 black dots and 10 white dots. Now the black and white dots represent tetrahedral numbers. Thus, the 5th square pyramidal number 55 is the sum of the 4th tetrahedral number, 20 and the 5th tetrahedral number, 35.

3. This activity gives students an opportunity to display creativity and through hands-on modeling achieve a better understanding of three dimensional figures.

4.
4th $10 + 20 = 30$
5th $20 + 35 = 55$
6th $35 + 56 = 91$
7th $56 + 84 = 140$
8th $84 + 120 = 204$
9th $120 + 165 = 285$

5. The formula for square pyramidal numbers is $\dfrac{n(n+1)(2n+1)}{6}$. The formula can be derived using the method of finite differences or by using the fact that a square pyramidal number is the sum of two tetrahedral numbers.

EXERCISE 69

1.
5th $40 + 35 = 75$
6th $75 + 51 = 126$
7th $126 + 70 = 196$
8th $196 + 92 = 288$
9th $288 + 117 = 405$
10th $405 + 145 = 550$
11th $550 + 176 = 726$
12th $726 + 210 = 936$

2. The nth pentagonal pyramidal number is $\dfrac{n^2(n+1)}{2}$.

EXERCISE 70

1. 5th $50 + 45 = 95$
    6th $95 + 66 = 161$
    7th $161 + 91 = 252$
    8th $252 + 120 = 372$
    9th $372 + 153 = 525$
   10th $525 + 190 = 715$
   11th $715 + 231 = 946$
   12th $946 + 276 = 1222$

2. The nth hexagonal pyramidal number is $\dfrac{n(n+1)(4n-1)}{6}$.

EXERCISE 71

1. 4th: $16 + 4(6) = 40$ 5th: $25 + 4(10) = 65$ 6th: $36 + 4(15) = 96$
   7th: $49 + 4(21) = 133$ 8th: $64 + 4(28) = 176$

2. 4th $30 + 40 = 70$
    5th $70 + 65 = 135$
    6th $135 + 96 = 231$
    7th $231 + 133 = 364$
    8th $364 + 176 = 540$
    9th $540 + 225 = 765$
   10th $765 + 280 = 1045$

3. 4 $40 = 16 + 4(6) = 70$
    5     $65 = 25 + 4(10) = 135$
    6     $96 = 36 + 4(15) = 231$
    7   $133 = 49 + 4(21) = 364$
    8   $176 = 64 + 4(28) = 540$

4. The nth 4-pointed star number is $n(3n-2)$. The nth 4-pointed star pyramidal number is $\dfrac{n(n+1)(2n-1)}{2}$.

EXERCISE 72

1. The formula here is $n(n+2)$, where $n = 1, 2, 3, \ldots$

3,   8,   15,   24,   35,   48,   63,   80,   99,
120,   143,   168,   195,   224,   255,   288,   323,   360,
399,   440,   483,   528,   575,   624,   675

2. a) 4, $4 + 6 = 10$, $4 + 6 + 8 = 18$, $4 + 6 + 8 + 10 = 28$
  b) By the formula $n(n+3)$, where $n = 1, 2, 3, \ldots$
  c) 4,   10,   18,   28,   40,   54,   70,   88,   108,
     130,   154,   180,   208,   238,   270,   304,   340,   378,
     418,   460,   504,   550,   598,   648,   700

3. a) 5, $5 + 7 = 12$, $5 + 7 + 9 = 21$, $5 + 7 + 9 + 11 = 32$

b) By the formula $n(n + 4)$, where $n = 1, 2, 3, \ldots$

c) 5,   12,   21,   32,   45,   60,   77,   96,   117,

140,   165,   192,   221,   252,   285,   320,   357,   396,

437,   480,   525,   572,   621,   672,   725

4.

2-type			3-type			4-type			5-type	
3	12	20	3	15	26	3	18	32	3	21	38
4	20	40	4	24	50	4	28	60	4	32	70
5	30	70	5	35	85	5	40	100	5	45	115
6	42	112	6	48	133	6	54	154	6	60	175

5. 2-type: $\dfrac{n(n + 1)(n + 2)}{3}$

3-type: $\dfrac{n(n + 1)(2n + 7)}{6}$

4-type: $\dfrac{n(n + 1)(n + 5)}{3}$

5-type: $\dfrac{n(n + 1)(2n + 13)}{6}$

EXERCISE 73

1. 27,   64,   125,   216,   343,   512,   729,   1000

EXERCISE 74

In problems 1, 3, and 4, the answer lists the numbers such that the sum of their cubes equals the given integer.

1. a) 1, 1, 1, 1, 1, 1, 2   b) 1, 1, 1, 1, 2, 2   c) 1, 1, 3   d) 2, 3

e) 2, 2, 3   f) 1, 1, 1, 4   g) 1, 1, 1, 1, 2, 4   h) 2, 3, 3, 3

2. 16, 28, 35, 54, 65, 72, 91

3. a) 1, 1, 2, 3, 3   b) 1, 2, 2, 3, 3, 3, 3   c) 3, 4, 5   d) 1, 1, 5, 6

e) 2, 2, 4, 6, 6   f) 1, 6, 8   g) 1, 3, 3, 6, 9   h) 3, 4, 6, 8, 8

i) 6, 8, 10   j) 1, 5, 7, 12

4. a) 18, 20 or 2, 24   b) 10, 27 or 19, 24   c) 18, 30 or 4, 32

d) 2, 34 or 15, 33   e) 9, 34 or 16, 33   f) 3, 36 or 27, 30

g) 17, 39 or 26, 36   h) 12, 40 or 31, 33   i) 4, 48 or 36, 40

5. $854 = 5^3 + 9^3$, $855 = 7^3 + 8^3$

6. $728 = 8^3 + 6^3$, $730 = 9^3 + 1^3$; another pair is $126 = 1^3 + 5^3$, $128 = 4^3 + 4^3$

EXERCISE 75

1. $(1, 8, 4, 9)$ $1^2 + 8^2 + 4^2 = 81 = 9^2$
2. $(2, 3, 6, 7)$ $2^2 + 3^2 + 6^2 = 49 = 7^2$
3. $(2, 4, 4, 6)$ $2^2 + 4^2 + 4^2 = 36 = 6^2$
4. $(2, 6, 9, 11)$ $2^2 + 6^2 + 9^2 = 121 = 11^2$

EXERCISE 76

1. a) $5 + 19$ b) $13 + 19$ c) $5 + 41$ d) $7 + 43$
   e) $7 + 61$ f) $41 + 59$ g) $37 + 73$ h) $73 + 181$
   i) $101 + 197$ j) $61 + 239$ k) $223 + 233$ l) $239 + 241$
   m) $229 + 283$ n) $257 + 293$ o) $293 + 307$

2. a) $7 + 19$ b) $7 + 31$ c) $3 + 41$ d) $5 + 47$
   e) $7 + 53$ f) $19 + 79$ g) $5 + 97$ h) $7 + 101$
   i) $5 + 229$ j) $19 + 283$ k) $3 + 419$ l) $5 + 491$
   m) $3 + 503$ n) $7 + 577$ o) $5 + 601$

3. a) $3 + 5 + 7$ b) $7 + 7 + 13$ c) $3 + 5 + 23$
   d) $3 + 19 + 31$ e) $11 + 29 + 29$ f) $11 + 29 + 61$
   g) $7 + 31 + 113$ h) $5 + 7 + 241$ i) $3 + 13 + 281$
   j) $3 + 5 + 293$ k) $47 + 73 + 313$ l) $79 + 101 + 307$
   m) $3 + 5 + 503$ n) $3 + 5 + 587$ o) $3 + 7 + 601$

4. The sum of two odd numbers is even. Thus, if the square is odd, one of the two primes must be 2. 11^2 and 17^2 cannot be expressed as 2 plus a prime.

5. $3 = 1 + 2 - 5 = 1 - 2 + 3 + 3$
   $7 = 1 - 2 + 3 + 5$
   $11 = 1 - 2 + 3 - 5 + 7 + 7$
   $13 = 1 + 2 - 3 - 5 + 7 + 11$
   $17 = 1 + 2 - 3 - 5 + 7 - 11 + 13 + 13$
   $19 = 1 - 2 + 3 + 5 - 7 - 11 + 13 + 17$
   $23 = 1 + 2 + 3 + 5 - 7 + 11 - 13 - 17 + 19 + 19$
   $29 = 1 + 2 + 3 + 5 + 7 + 11 - 13 + 17 + 19 - 23$

EXERCISE 77

1. 12: $1 + 11, 3 + 9, 5 + 7$; 3 ways
   14: $1 + 13, 3 + 11, 5 + 9, 7 + 7$; 4 ways
   16: $1 + 15, 3 + 13, 5 + 11, 7 + 9$; 4 ways
   18: $1 + 17, 3 + 15, 5 + 13, 7 + 11, 9 + 9$; 5 ways
   20: $1 + 19, 3 + 17, 5 + 15, 7 + 13, 9 + 11$; 5 ways
   22: $1 + 21, 3 + 19, 5 + 17, 7 + 15, 9 + 13, 11 + 11$; 6 ways

24: $1 + 23, 3 + 21, 5 + 19, 7 + 17, 9 + 15, 11 + 13$; 6 ways
26: $1 + 25, 3 + 23, 5 + 21, 7 + 19, 9 + 17, 11 + 15, 13 + 13$; 7 ways
28: $1 + 27, 3 + 25, 5 + 23, 7 + 21, 9 + 19, 11 + 17, 13 + 15$; 7 ways
30: $1 + 29, 3 + 27, 5 + 25, 7 + 23, 9 + 21, 11 + 19, 13 + 17, 15 + 15$; 8 ways
32: $1 + 31, 3 + 29, 5 + 27, 7 + 25, 9 + 23, 11 + 21, 13 + 19, 15 + 17$; 8 ways
34: $1 + 33, 3 + 31, 5 + 29, \ldots, 13 + 21, 15 + 19, 17 + 17$; 9 ways
36: $1 + 35, 3 + 33, 5 + 31, \ldots, 13 + 23, 15 + 21, 17 + 19$; 9 ways
38: $1 + 37, 3 + 35, 5 + 33, \ldots, 15 + 23, 17 + 21, 19 + 19$; 10 ways
40: $1 + 39, 3 + 37, 5 + 35, \ldots, 15 + 25, 17 + 23, 19 + 21$; 10 ways
42: $1 + 41, 3 + 39, 5 + 37, \ldots, 17 + 25, 19 + 23, 21 + 21$; 11 ways
44: $1 + 43, 3 + 41, 5 + 39, \ldots, 17 + 27, 19 + 25, 21 + 23$; 11 ways
46: $1 + 45, 3 + 43, 5 + 41, \ldots, 19 + 27, 21 + 25, 23 + 23$; 12 ways

2. Divide the number by 4. If the result is a whole number, that is the number of ways. Otherwise, add $1/2$ to your quotient with remainder and that will be the number of ways.
    $32 \div 4 = 8$, so 32 can be written in 8 ways.
    $38 \div 4 = 9\frac{1}{2}$, $9\frac{1}{2} + \frac{1}{2} = 10$, so 38 can be written in 10 ways.

3. See the answer to 1.
    12: 3 ways   14: 3 ways   16: 4 ways   18: 4 ways   20: 5 ways
    22: 5 ways   24: 6 ways   26: 6 ways   28: 7 ways   30: 7 ways

4. Divide the even number by 4 and the quotient is the number of ways.
    $12 \div 4 = 3$, so 12 can be written in 3 ways.
    $14 \div 4 = 3\frac{1}{2}$, and the quotient is three. So, 14 can be written in 3 ways.

5, 6. The number of ways, N, to express the odd integer n as a sum of three odd integers is given in the table

n	11	13	15	17	19	21	23	25	27	29	31	33	35	37	39
N	4	5	7	8	10	12	14	16	19	21	24	27	30	33	37

It is possible to create a formula using the greatest integer function.

$$N = \sum_{i=1} \left\{ \frac{1}{2}(n + 3 - 4i) - \left[\frac{n - (2i - 1)}{4} \right] \right\}$$

where N is the number of ways and n is the given odd number, and i goes from 1 up to but not including the value of i that makes the quantity in the braces negative. The brackets in $\left[\frac{n - (2i - 1)}{4} \right]$ signify the greatest integer function.

7. 17: $1 + 3 + 13, 1 + 5 + 11, 1 + 7 + 9, 3 + 5 + 9$; 4 ways
    19: $1 + 3 + 15, 1 + 5 + 13, 1 + 7 + 11, 3 + 5 + 11, 3 + 7 + 9$; 5 ways
    21: $1 + 3 + 17, 1 + 5 + 15, 1 + 7 + 13, 1 + 9 + 11, 3 + 5 + 13,$
        $3 + 7 + 11, 5 + 7 + 9$; 7 ways
    23: $1 + 3 + 19, 1 + 5 + 17, 1 + 7 + 15, 1 + 9 + 13, 3 + 5 + 15,$
        $3 + 7 + 13, 3 + 9 + 11, 5 + 7 + 11$; 8 ways

25: $1 + 3 + 21, 1 + 5 + 19, 1 + 7 + 17, 1 + 9 + 15, 1 + 11 + 13$,
   $3 + 5 + 17, 3 + 7 + 15, 3 + 9 + 13, 5 + 7 + 13, 5 + 9 + 11$; 10 ways
27: 12 ways   29: 14 ways   31: 16 ways   33: 19 ways   35: 21 ways

8. The following formula will give the number of ways, N, for an odd integer to be written as a sum of three *distinct* odds.

$$N = \sum_{i=1}^{} \left\{ \frac{1}{2}(n + 1 - 4i) - \left[\frac{n - (2i - 3)}{4} \right] \right\}$$

where N is the number of ways and n is the given odd number, and i goes from 1 up to but not including the value of i that makes the quantity in the braces negative. The brackets in $\left[\dfrac{n - (2i - 3)}{4} \right]$ signify the greatest integer function.

EXERCISE 78

1. 17: $8 + 9$ 18: $9 + 9, 8 + 10, 4 + 14, 6 + 12$
   19: $4 + 15, 9 + 10$ 20: $4 + 16, 6 + 14, 8 + 12, 10 + 10$
   21: $6 + 15, 9 + 12$ 22: $4 + 18, 6 + 16, 8 + 14, 10 + 12$
   23: $8 + 15, 9 + 14$
   24: $4 + 20, 6 + 18, 8 + 16, 9 + 15, 10 + 14, 12 + 12$
   25: $4 + 21, 9 + 16, 10 + 15$
   26: $4 + 22, 6 + 20, 8 + 18, 10 + 16, 12 + 14$

EXERCISE 79

1. a)   A   1   2   3   4
         B   4   3   2   1
   b)   A   1   2   3   4   5
         B   1   5   4   3   2
   c)   A   1   2   3   4   5   6
         B   6   5   4   3   2   1
   d)   A   1   2   3   4   5   6   7
         B   1   5   4   3   2   7   6
   e)   A   1   2   3   4   5   6   7   8
         B   4   5   8   1   2   7   6   3
   f)   A   1   2   3   4   5   6   7   8   9
         B   1   9   4   3   8   7   6   5   2
   g)   A   1   2   3   4   5   6   7   8   9   10
         B   10   9   8   7   6   5   4   3   2   1
   h)   A   1   2   3   4   5   6   7   8   9   10   11
         B   1   11   10   9   8   7   6   5   4   3   2
   i)   A   1   2   3   4   5   6   7   8   9   10   11   12
         B   12   11   10   9   8   7   6   5   4   3   2   1

2. 1) If n is 1 less than a prime (and thus $n+1$ is a prime), use the matching pattern

1	2	3	4		n
$+n$	$+n-1$	$+n-2$	$+n-3$	$\cdots$	$+1$
$\overline{n+1}$	$\overline{n+1}$	$\overline{n+1}$	$\overline{n+1}$		$\overline{n+1}$

2) We assume that all the numbers 1 through $(n-1)$ are prime pair numbers. Show n is a prime pair number. If n is not 1 less than a prime (and thus, $n+1$ is not a prime), start with 1 and find the first number k in the sequence $1, 2, 3, \ldots$, such that $(n+k)$ is a prime. Then use the matching pattern.

1	2	3	$\ldots$	$k-1$	k	$k+1$	$k+2$	$\cdots$	n
					$+n$	$+n-1$	$+n-2$		$+k$
					$\overline{n+k}$	$\overline{n+k}$	$\overline{n+k}$	$\ldots$	$\overline{n+k}$

Since $(k-1) < n$, the number $(k-1)$ has already been shown to be a prime pair number and the matching is known. Thus, the problem is solved.

EXERCISE 80

1, 2. Notice that a prime line may not be a prime circle, but every prime circle is also a prime line. Each number from 2 through 20 is a prime line number. The even numbers are also prime circle numbers. The arrangements are shown below.

```
2:  1 2 ♦ circle
3:  3 2 1
4:  4 3 2 1 ♦ circle
5:  5 2 3 4 1
6:  6 1 4 3 2 5 ♦ circle
7:  7 6 1 4 3 2 5
8:  6 1 2 5 8 3 4 7 ♦ circle
9:  7 6 1 4 3 2 5 8 9
10: 7 6 1 4 3 2 5 8 9 10 ♦ circle
11: 11 2 5 6 7 10 1 4 9 8 3
12: 12 5 6 7 10 1 4 9 8 3 2 11 ♦ circle
13: 1 12 7 10 13 6 11 2 9 4 3 8 5
14: 14 3 4 1 12 7 10 13 6 11 2 9 8 5 ♦ circle
15: 15 4 1 12 7 10 13 6 11 2 9 14 3 8 5
16: 14 3 16 15 4 1 12 7 10 13 6 11 2 9 8 5 ♦ circle
17: 3 16 15 4 1 12 7 10 13 6 11 2 17 14 9 8 5
18: 5 18 11 6 13 10 7 12 1 4 15 16 3 2 17 14 9 8 ♦ circle
19: 3 16 15 4 1 12 7 10 13 6 11 2 17 14 9 8 5 18 19
20: 3 2 1 4 7 6 5 8 9 10 13 16 15 14 17 12 19 18 11 20 ♦ circle
```

3. It appears that no odd collection of consecutive integers can be a circle, thus, in 1 2 3 4 5 there are 3 odds and only 2 evens. Since we must have odds and evens alternate to get primes (odds), then if we start with an odd, we end with an odd and the 1st and last numbers will be odds (whose sum is even and so not a prime).

If we start with an even, then at the end we have 2 odds (again the sum will be even). Thus, only an even collection of consecutive integers can be a circle, and it appears that every even collection can be made into a circle.

4. a, b, c. Answers will vary.

```
               ▲
              1  2
             1  2  3
            1  2  3  4
           1  4  3  2  5
          1  4  3  2  5  6
         1  4  3  2  5  6  7
        1  2  5  6  7  4  3  8
       1  2  5  6  7  4  3  8  9
      1  2  5  6  7  4  3  8  9  10
     1  2  9  10 7  4  3  8  5  6  11
    1  2  9  10 7  4  3  8  5  6  11 12
   1  12 11 6  7  4  3  2  5  8  9  10 13
  1  2  3  8  5  12 11 6  7  4  13 10 9  14
 1  2  3  8  5  12 11 6  7  4  13 10 9  14 15
1  2  3  8  5  12 11 6  7  4  13 10 9  14 15 16
 1  16 15 14 9  10 13 4  7  6  11 12 5  8  3  2 17
  1  4  3  2  17 12 5  8  11 6  7  10 9  14 15 16 13 18
   1  4  3  2  17 12 5  8  11 6  7  10 9  14 15 16 13 18 19
    1  4  3  2  17 12 5  8  11 6  7  10 19 18 13 16 15 14 9  20
```

EXERCISE 81

1. 2, 3, 8, 10, 24, 48, 80, 82, 168

2. No. 3 is the only beprisque number that is odd because 2 is the only even prime.

3.

$2 = 2$	$3 = 3$	$4 = 2 + 2$	$5 = 2 + 3$
$6 = 3 + 3$	$7 = 2 + 2 + 3$	$8 = 8$	$9 = 3 + 3 + 3$
$10 = 10$	$11 = 3 + 8$	$12 = 2 + 10$	$13 = 3 + 10$
$14 = 2 + 2 + 10$	$15 = 2 + 3 + 10$	$16 = 8 + 8$	

4. Answers will vary.

EXERCISE 82

1. 11, 23, 29, 41, 53, 83, 89

2. a) 131   b) 113   c) There are no Germain primes that end in 7 since $(2 \times 7) + 1$ ends in 5. d) 179

3. 11, 23 and 83 have a remainder of 3 when divided by 4. The others have a remainder of 1.

4. In the 3 cases above with a remainder of 3, $2p + 1$ divides $2^p - 1$.
That is, $2^{11} - 1/23 = 89$; $2^{23} - 1/47 = 178481$;
$2^{83} - 1/167 = 5791261411327564 9087721$.

5. A likely choice: $89 \rightarrow 179 \rightarrow 359 \rightarrow 719 \rightarrow 1439$

EXERCISE 83

1. (3, 5)   (5, 7)   (11, 13)   (17, 19)   (29, 31)   (41, 43)   (59, 61)   (71, 73)

2. The sums of twin primes (excluding 3, 5) are multiples of 12.

3. The sum of p and $p + 2$ is a multiple of 12 because:

$-(p + 1)$ is between p, $p + 2$ and is therefore even.

$-(p + 1)$ must also be a multiple of three, since in every set of three consecutive numbers there is a multiple of three and p and $p + 2$ are prime.

Thus, $p + (p + 2) = 2(p + 1) = 2 \times 2 \times 3 \times n$, a multiple of 12.

4. The product of twin primes is one less than the square of the number between the primes.

5. $p(p + 2) = p^2 + 2p, (p + 1)^2 = p^2 + 2p + 1.$

6. Answers will vary.

a) $5 - 3, 7 - 5, 13 - 11, 19 - 17, 43 - 41, 73 - 71, 103 - 101, 109 - 107,$
$139 - 137, 151 - 149$

b) $7 - 3, 11 - 7, 17 - 13, 23 - 19, 41 - 37, 47 - 43, 71 - 67, 83 - 79,$
$101 - 97, 107 - 103$

c) $11 - 5, 13 - 7, 17 - 11, 19 - 13, 23 - 17, 29 - 23, 37 - 31, 43 - 37, 47 - 41,$
$53 - 47$

EXERCISE 84

1. $28 = 22 + 6$ $29 = 15 + 14$ $30 = 15 + 15$ $31 = 21 + 10$
$32 = 26 + 6$ $33 = 21 + 6 + 6$ $34 = 22 + 6 + 6$ $35 = 21 + 14$
$36 = 22 + 14$ $37 = 22 + 15$ $38 = 26 + 6 + 6$ $39 = 33 + 6$
$40 = 34 + 6$ $41 = 26 + 15$ $42 = 22 + 14 + 6$ $43 = 33 + 10$
$44 = 34 + 10$ $45 = 15 + 15 + 15$ $46 = 21 + 15 + 10$ $47 = 33 + 14$
$48 = 33 + 15$

2. $86 = 65 + 21 = 51 + 35$ $106 = 91 + 15 = 85 + 21$
$146 = 111 + 35 = 95 + 51$ $155 = 145 + 10 = 141 + 14$
$185 = 159 + 26 = 146 + 39$

3. 30, 42, 66, 70, 78

4. a) $\{1, 3\}, 1, 2$ b) $\{1, 5\}, 1, 2$ c) $\{1, 2, 3, 6\}, 2, 4$
d) $\{1, 2, 7, 14\}, 2, 4$ e) $\{1, 3, 5, 15\}, 2, 4$ f) $\{1, 23\}, 1, 2$
g) $\{1, 2, 3, 5, 6, 10, 15, 30\}, 3, 8$ h) $\{1, 31\}, 1, 2$ i) $\{1, 3, 11, 33\}, 2, 4$
j) $\{1, 2, 3, 6, 7, 14, 21, 42\}, 3, 8$ k) $\{1, 59\}, 1, 2$
l) $\{1, 2, 3, 6, 11, 22, 33, 66\}, 3, 8$

5. The number of divisors equals 2 raised to the number of distinct primes in the prime factorization.

EXERCISE 85

1–7. The solutions were found using a table of primes with a six-digit limit. Many answers are possible.

$1 - 13 - 139 - 1399 - 13999 - 139999$
$1 - 19 - 197 - 1979 - 19793 - 197933$

2 – 23 – 233 – 2333 – 23333
2 – 29 – 293 – 2939 – 29399 – 293999
3 – 37 – 373 – 3733 – 37337 – 373379
5 – 59 – 593 – 5939 – 59393 – 593933
7 – 71 – 719 – 7193 – 71933 – 719333

EXERCISE 86

1. a) $3 + 13$ b) $3 + 15$ c) $7 + 13$ d) $9 + 13$ e) $3 + 21$
f) $1 + 31$ g) $15 + 33$ h) $7 + 49$ i) $1 + 63$ j) $1 + 69$
k) $3 + 79$ l) $3 + 93$ m) $25 + 79$ n) $1 + 211$ o) $15 + 303$

2. a) $1 + 25$ b) $3 + 25$ c) $9 + 21$ d) $1 + 33$ e) $3 + 33$
f) $1 + 37$ g) $3 + 43$ h) $1 + 51$ i) $9 + 51$ j) $3 + 75$
k) $1 + 87$ l) $1 + 93$ m) $3 + 99$ n) $15 + 201$ o) $3 + 297$

3. a) $3 + 3 + 7$ b) $1 + 7 + 7$ c) $1 + 3 + 13$ d) $1 + 3 + 15$
e) $3 + 3 + 15$ f) $1 + 7 + 25$ g) $1 + 7 + 37$ h) $3 + 3 + 51$
i) $3 + 7 + 51$ j) $3 + 7 + 69$ k) $3 + 3 + 75$ l) $3 + 3 + 93$
m) $1 + 3 + 99$ n) $1 + 3 + 211$ o) $1 + 3 + 303$

4. Answers will vary.
a) Except for 2, all primes are odd. *All* luckies are odd.
b) Distribution of luckies and primes is similar.
c) There are twin primes, that is, primes separated by 1 number. For example $(3, 5), (5, 7), (11, 13)$. There are also twin luckies $(1, 3), (7, 9), (13, 15)$. There are 33 twin luckies < 1000, and 35 twin primes < 1000.
d) Sum of twin primes and also twin luckies is divisible by 4.
e) Product of twin luckies and also twin primes is 1 less than the square of the number between the twins.

EXERCISE 87

1.   a) 5 odd, 4 even   b) 7 odd, 3 even   c) 5 odd, 5 even   d) 3 odd, 7 even
e) 8 odd, 2 even   f) 3 odd, 7 even   g) 6 odd, 4 even   h) 8 odd, 2 even
i) 2 odd, 8 even   j) 4 odd, 6 even

2.   a) 5 odd, 4 even   b) 12 odd, 7 even   c) 17 odd, 12 even
d) 20 odd, 19 even   e) 28 odd, 21 even   f) 31 odd, 28 even
g) 37 odd, 32 even   h) 45 odd, 34 even   i) 47 odd, 42 even
j) 51 odd, 48 even

EXERCISE 88

1. There is one single-digit balanced number, 2.
2. a) Yes: 6 prime, 6 composite   b) No: 6 prime, 8 composite   c) No: 7 prime, 9 composite

3. The number of balanced numbers is not unlimited because once the number of composites exceeds the number of primes, namely after 14, the equilibrium is skewed and cannot be regained.

EXERCISE 89

1. $F_2 = 2^{2^2} + 1 = 2^4 + 1 = 17$ prime

$F_3 = 2^{2^3} + 1 = 2^8 + 1 = 257$ prime

$F_4 = 2^{2^4} + 1 = 2^{16} + 1 = 65,537$ prime, the largest known Fermat prime.

Since the time of Euler who in 1732 factored $F_5 = 641 \times 6,700,417$, researchers have been at work locating factors of Fermat numbers. The results are maintained at www.prothsearch.net/Fermat.html and many details are kept up-to-date at www.primes.utm.edu.

2. Regular polygons with 3, 4, 5, 6, 8, 10, 12, 15, 16, 17, 20, 24, 30, ... sides are constructible. There are 24 constructible polygons (odd- or even-sided) less than 100, 37 with sides not exceeding 300, 52 with sides not exceeding 1000, and 206 with sides not exceeding 1,000,000.

Theoretically, polygons with the number of sides shown above can be constructed by straightedge and compass; the actual carrying out of the construction is not easy.

For the 17-sided polygon, the construction is rather involved, and many pages are required for the exposition of a 257-gon. Professor Hermes of Lingen devoted ten years of his life to the regular polygon of 65,537 sides!

EXERCISE 90

1. 1: 3   2: 9   3: 25   4: 65   5: 161   6: 385

     7: 897   8: 2049   9: 4609   10: 10241   11: 22529

     12: 49153   13: 106497   14: 229377   15: 491521

2. 5: 159   10: 10239   15: 491519   20: 20971519   25: 838860799

3. Answers will vary. Some possibilities: Cullen numbers are odd. The first three Cullen numbers in (1) are squares. The two types of Cullen numbers for each n differ by 2.

EXERCISE 91

In order to answer the questions in this exercise develop a list of the first 100 integers with their corresponding prime factorizations and $S(n)$.

1. (5, 6)   (8, 9)   (15, 16)   (77, 78)

2. a) (2, 3)   (3, 4)   (4, 5)   (14, 15)   (20, 21)   (24, 25)   (63, 64)

     (80, 81)   (98, 99)

     b) (6, 7)   (7, 8)   (9, 10)   (27, 28)   (35, 36)   (49, 50)   (65, 66)

     c) (21, 22)   (48, 49)   (55, 56)   (99, 100)

     d) (10, 11)   (11, 12)   (32, 33)   (44, 45)

3. 97, the largest prime

4. $S(n) = 13$ occurs 8 times.

EXERCISE 92

1. 1

```
1   2   1
1   2   3   2   1
1   2   3   4   3   2   1
1   2   3   4   5   4   3   2   1
1   2   3   4   5   6   5   4   3   2   1
1   2   3   4   5   6   7   6   5   4   3   2   1
1   2   3   4   5   6   7   8   7   6   5   4   3   2   1
1   2   3   4   5   6   7   8   9   8   7   6   5   4   3   2   1
1   2   3   4   5   6   7   9   0   0   9   8   7   6   5   4   3   2   1
```
                          (last line not palindromic as others are)

2. a) 111,111   b) 11,111,111   c) 1,111,111,111   d) 111,111,111,111

3. a) 111(1001)   b) 1111(10001)   c) 111(1001001)   d) 11111(100001)

4. See problem 2. $111,111 = 10101 \times 11 = 3(7)(13)(37)(11)$
    $11,111,111 = 1010101 \times 11 = 73(101)(137)(11)$

5. 13 divides 12 1s; quotient is 8,547,008,547
    17 divides 16 1s; quotient is 65,359,477,124,183

EXERCISE 93

1. 2 3 4 2 1 3 1 4

2. 2 3 7 2 6 3 5 1 4 1 7 6 5 4

3. 3 6 8 1 3 1 5 7 6 4 2 8 5 2 4 7

EXERCISE 94

1. a) 2   b) 4   c) 6   d) 8   e) 10   f) 12   g) 14   h) 16
    i) 18   j) 11   k) 13   l) 15   m) 17   n) 103   o) 105

2. a, b, d, g, i, j, l, n, and p are social

EXERCISE 95

1. b, c, and e are divisible by 3.

2. b, c, and g are divisible by 9.

3. A number, $abcd$, in expanded notation reads $1000a + 100b + 10c + d$. This expression can also be written as $999a + a + 99b + b + 9c + c + d$. Since we know $999a + 99b + 9c$ is divisible by both 9 and 3, the number $abcd$ is divisible by 9 (or 3) if the sum of $a + b + c + d$ is a multiple of 9 (or 3). This argument can be modified to cover an arbitrary number.

EXERCISE 96

1. a, c, e, g, h, i, j

2. Answers will vary.

2: 110;   3: 111;   4: 112;   5: 140;   6: 116;   7: 133;   8: 152;
9: 108;   10: 190;   11: 209;   12: 156;   13: 247;   14: 266;   15: 195;
16: 448;   17: 476;   18: 288;   19: 874

3. Answers will vary.

a) 3, 12, 21   b) 5, 50, 140   c) 7, 70, 133
d) 9, 18, 27   e) 209, 308, 407   f) 247, 364, 481
g) 266, 392, 518   h) 195, 285, 375   i) 448, 592, 736
j) 476, 629, 782   k) 198, 288, 378   l) 874, 1387, 1558
m) 399, 588, 777   n) 2398, 2596, 2794   o) 4975, 5875, 7675
p) 17887, 25699, 25978   q) 37999, 38998, 39997   r) 2599960, 2688880,
                                                                        2698960

4. All are additive multidigital numbers.

$5! = 120$, divisor 3 $6! = 720$, divisor 9
$7! = 5040$, divisor 9 $8! = 40320$, divisor 9
$9! = 362880$, divisor 27 $10! = 3628800$, divisor 27
$11! = 39916800$, divisor 36 $12! = 479001600$, divisor 27

Refer to D. R. Kaprekar. "On Kaprekar's Harshad Numbers." *Journal of Recreational Mathematics*, Vol. 13: 1 (1980–81), pp. 2–3.

EXERCISE 97

1. *a*, *b*, *d*, *e*, *g*, and *i* are multiplicative multidigital numbers.

2. No. 48 is a multiple of 12. $48 \div (4+8) = 4$, but 48 is *not* divisible by the product $4 \times 8 = 32$. Thus, 48 is an additive but not a multiplicative multidigital number.

EXERCISE 98

1. a) 7   b) 3   c) 4   d) 6   e) 6   f) 2   g) 5   h) 1   i) 1   j) 7

2. Answers will vary.

Refer to D. R. Kaprekar. "An Interesting Property of the Number 6174." *Scripta Mathematica*, Vol. 21 (1955), p. 304.

EXERCISE 99

1, 2. In verifying remember: given a number, subtract the smaller from the larger.

3. For 4-digit and 5-digit numbers, follow the same rules as for 3-digit numbers: not all the digits are the same, no palindromes, subtract the smaller from the larger.

5342	7154	4321	8765
− 2435	− 4517	− 1234	− 5678
2907	2637	3087	3087
+ 7092	+ 7362	+ 7803	+ 7803
9999	9999	10890	10890

There is more than one pattern for 4-digit numbers.

65172	43312	54321	37132
− 27156	− 21334	− 12345	− 23173
38016	21978	41976	13959
+ 61083	+ 87912	+ 67914	+ 95931
99099	109890	109890	109890

There is more than one pattern for 5-digit numbers.

EXERCISE 100

1. *c, d, f, j, m, o*

2. Primes

3. The number of doubling numbers is *probably* unlimited. Although not all primes are doubling numbers, those that have been found are primes and there is an unlimited number of primes.

For more information on doubling numbers, see *CMC Communicator, Newsletter of the California Mathematics Council,* Vol. 4: 3 (September 1978).

EXERCISE 101

1. Good numbers and their partitions:

    1: (1)   4: (2, 2)   9: (3, 3, 3)   10: (2, 4, 4)
    11: (2, 3, 6)   16: (4, 4, 4, 4) 17: (4, 4, 3, 6)
    18: (3, 3, 6, 6) 20: (2, 6, 6, 6)

For more information on good numbers, see "News and Letters." *Mathematics Magazine,* Vol. 51: 3 (May 1978), p. 205.

EXERCISE 102

1. All given semiperfect numbers are nearly good. The partitions and the denominators of the unit fractions that sum to 1 are shown.

a) $1 + 4 + 5 + 10$, (20, 5, 4, 2)

b) $4 + 8 + 12$, (6, 3, 2)

c) $1 + 2 + 3 + 4 + 6 + 8$, (24, 12, 8, 6, 4, 3)

d) $2 + 3 + 10 + 15$, (15, 10, 3, 2)

e) $2 + 8 + 10 + 20$, (20, 5, 4, 2)

f) $8 + 16 + 24$, (6, 3, 2)

g) $3 + 6 + 18 + 27$, (18, 9, 3, 2)

h) $4 + 6 + 20 + 30$, (15, 10, 3, 2) or $3 + 12 + 15 + 30$, (20, 5, 4, 2)

i) $11 + 22 + 33$, (6, 3, 2)

j) $1 + 3 + 8 + 24 + 36$, (72, 24, 9, 3, 2)

2. All semiperfect numbers are nearly good. If N is semiperfect then it has a set of proper divisors $d_1, d_2, \ldots, d_k$ such that $d_1 + d_2 + \cdots + d_k = N$. (1) Since each d_i divides N, let $\dfrac{N}{d_i} = a_i$ then $\dfrac{d_i}{N} = \dfrac{1}{a_i}$. Then after dividing (1) by N, $\dfrac{d_1}{N} + \dfrac{d_2}{N} + \cdots + \dfrac{d_k}{N} = \dfrac{N}{N}$ the result is $\dfrac{1}{a_1} + \dfrac{1}{a_2} + \cdots + \dfrac{1}{a_k} = 1$.

EXERCISE 103

1. $135 = 1^1 + 3^2 + 5^3$ $153 = 1^1 + 5^3 + 3^3$ $175 = 11 + 7^2 + 5^3$
$224 = 2^5 + 2^7 + 4^3$ $226 = 2^1 + 2^3 + 6^3$

2. $2, 3, 5, 7, 43 = 4^2 + 3^3$

See J. Randle. "Problem Note 3208 Powerful Numbers", *The Mathematical Gazette*, Vol. 52, (December 1968), p. 383 and S. Kahan. "Powerful Integers." *Journal of Recreational Mathematics*, Vol. 10: 3 (1977–1978), p. 209.

EXERCISE 104

1. a) $370, 371, 407$ b) $1634, 8208, 9474$ c) $54748, 92727, 93084$ d) 548834

2. No, because for no positive integers x, y do
we have $x^2 + y^2 = 10x + y$.

3. a) $160 - 217 - 352 - 160$ d) $919 - 1459 - 919$

4. b) $2178 - 6514 - 2178$

5. $1138 - 4179 - 9219 - 13139 - 6725 - 4338 - 4514 - 1138$

6. $2178 - 6514 - 2178$

For further information on amicable digital invariant pair numbers, see Madachy, J. S. **Mathematics on Vacation,** New York: Charles Scribner's Sons, 1966, pp. 163–165.

EXERCISE 105

1. $(2 + \sqrt{4})!$ **2.** $2^0 \times 5^2$ **3.** $(\sqrt{2+7})^3$ **4.** $3! \times 6$
5. $(3!)^2 + \sqrt{9}$ **6.** $4! \times \sqrt[3]{8}$ **7.** $(6 - \sqrt{4})^3$ **8.** $8^1 + 9^2$

EXERCISE 106

1. a) 8   b) 5   c) 7   d) 3   e) 9   f) 4   g) 2   h) 4   i) 4   j) 1

2. Answers will vary.

1	2	3	4	5	6	7	8	9
208	272	300	400	500	600	700	800	900
217	209	201	301	347	276	394	386	495
325	317	327	436	446	492	484	467	558
433	443	435	553	554	555	880	557	882

3. a) No   b) No   c) No   d) No   e) Perhaps   f) Perhaps
4. The additive digital roots of the triangular numbers form a repeating pattern nine
digits long, namely : 1, 3, 6, 1, 6, 3, 1, 9, 9, . . .

EXERCISE 107

1. a) 2   b) 1   c) 2   d) 1   e) 1   f) 2   g) 2   h) 2
i) 2   j) 3   k) 3   l) 2   m) 2   n) 2   o) 3
2. 39999 → 39 → 12 → 3
69999999 → 69 → 15 → 6
3. a) 10   b) 19   c) 199   d) 1999 . . . 9 (22 of the 9s)
4. There is no such integer.

EXERCISE 108

1. a) 0   b) 2   c) 8   d) 6   e) 8   f) 2   g) 0   h) 8   i) 6   j) 0
2. Answers will vary.

0	1	2	3	4	5	6	7	8	9
665	1111	211	311	411	511	611	711	811	911
588	11111	216	3111	436	5111	668	7111	1214	313
252	111111	621	1113	346	1511	821	1711	2221	9111
345	1111111	666	1311	364	1151	288	1171	3321	3311
201	11111111	673	1131	634	1115	1288	1117	8111	31113

EXERCISE 109

1. a) 1   b) 1   c) 2   d) 2   e) 1   f) 2   g) 3   h) 1
i) 2   j) 3   k) 3   l) 3   m) 2   n) 3   o) 3
2. Answers will vary.
49 → 36 → 18 → 8   55 → 25 → 10 → 0
3. a) 10   b) 25   c) 39   d) 77
4. The smallest number with a multiplicative persistence of 5 is 679.
5. Four digit numbers can have a multiplicative persistence of 1, 2, 3, 4, 5, and 6.
The smallest number with a multiplicative persistence of 7 is 68889.

EXERCISE 110

1. 13, 19, 23, 26, 29, 39, 46, 49, 59, 69, 79, 89
2. a) Section 103 as 1 | 03. Then 103/03 = 34 R 1
b) 206 = 2 × 103,   309 = 3 × 103
c) Yes, 206 and 309 are also modest numbers.

Section 206 as 2 | 06, so 206/06 = 34 R2.

Section 309 as 3 | 09, so 309/09 = 34 R3.

3. Answers will vary. The following numbers are modest:

a) 266, 399   b) 406, 609   c) 422, 633   d) 866, 1299

4. a) a | bcd;   ab | cd provided ab < cd   b) a | bcde; ab | cde

5. a) 2333 is modest. 2|333 implies 2333/333 = 7 R 2

b) not modest

c) 2666 is modest. 2|666 implies 2666/666 = 4 R 2

d) not modest

6. a) A number a333 is extremely modest when a = 1, 2.

b) A number a666 is extremely modest when a = 2, 3, 4.

c) A number a999 is extremely modest for any digit a = 1, 2, . . . , 8.

EXERCISE 111

1. a, c, d, f, g, k, and l are visible factor numbers.

2. a, c, d, j

EXERCISE 112

1. 10, 11, 12, 15, 20, 22, 24, 30, 33, 36, 40, 44, 48, 50, 55, 60, 66, 70, 77, 80, 88, 90, 99

2. *cab* may be nude. If $N = cab$ is to be nude then the nonzero digits amongst c, a, b must divide *cab* and at least one other number, not N, but formed from digits of N,must divide *cab*. The possibilities for this number are *ac, ca, ab, ba, bc, cb*. 20 is nude and 420 is nude. 24 is nude, but 324 is not nude.

3. $N = abab$ is nude since it is given that a, b divide ab. Thus, a, b divide $abab = ab$ (101) and also ab divides $abab$.

4. a) 132 (divisible by 1, 2, 3, 12)   312 (divisible by 1, 2, 3, 12, 13)

b) 936 (divisible by 3, 6, 9, 36, 39)   396 (divisible by 3, 6, 9, 36)

c) no permutations of 428 are nude.

d) 735 (divisible by 3, 5, 7, 35)

5. Answers will vary. 1212, 2424, 3636 are nude.

6. Answers will vary. See (4) above.

7. Answers will vary. 3120, 9360 are nude.

8. 27216 has 60 divisors including 1, 2, 6, 7, 12, 16, 21, 27, 72 and 216.

9. All nude numbers are visible factor numbers but not conversely.

EXERCISE 113

1. To find the number of terms in a consecutive number sequence, subtract the 1st term from the last term and then add 1. To find the sum, multiply the number of terms by the average of the first and last terms.

a) 42   b) 70   c) 49   d) 51   e) 99   f) 148   g) 225   h) 275

i) 366   j) 840

2. a) 21   b) 37   c) 59   d) 83   e) 201   f) 309   g) 2003   h) 9019
i) 13579

3. Yes. The sum is of an even and odd integer. That is, $(2n) + (2n - 1) = 4n - 1$ which is odd.

4. a) $5 + 6$ b) $9 + 10$ c) $18 + 19$ d) $24 + 25$ e) $26 + 27$
f) $32 + 33$ g) $40 + 41$ h) $54 + 55$

5. $2n - 1 = n + (n - 1)$

6. a) 33   b) 45   c) 63   d) 129   e) 276
f) 303   g) 612   h) 1509   i) 2970

7. The sum equals 3 times the middle integer. Thus, $10 + 11 + 12 = 3(11) = 33$.

8. Yes, because $n + (n + 1) + (n + 2) = 3n + 3 = 3(n + 1)$ which is the $(n + 1)$ — multiple of 3.

EXERCISE 114

1. a) $9 + 10 + 11$ b) $13 + 14 + 15$ c) $19 + 20 + 21$
d) $28 + 29 + 30$ e) $30 + 31 + 32$ f) $32 + 33 + 34$
g) $36 + 37 + 38$ h) $71 + 72 + 73$ i) $110 + 111 + 112$

2. a) $5 + 7 + 9$ b) $16 + 18 + 20$ c) $25 + 27 + 29$
d) $29 + 31 + 33$ e) $32 + 34 + 36$ f) $34 + 36 + 38$
g) $36 + 38 + 40$ h) $63 + 65 + 67$ i) $79 + 81 + 83$

3. a) 26   b) 46   c) 54   d) 78   e) 94   f) 126   g) 166   h) 210

4. The sum equals twice the sum of the first and last numbers.
$n + (n + 1) + (n + 2) + (n + 3) = 4n + 6 = 2(2n + 3) = 2(n + (n + 3))$.

5. Yes, because the sum of four consecutive integers is $2(2n + 3)$ and $(2n + 3)$ is an odd number.

6. a) $7 + 8 + 9 + 10$ b) $9 + \cdots + 12$ c) $11 + \cdots + 14$ d) $14 + \cdots + 17$
e) $18 + \cdots + 21$ f) $24 + \cdots + 27$ g) $30 + \cdots + 33$ h) $34 + \cdots + 37$
i) $36 + \cdots + 39$

7. a) 30   b) 35   c) 40   d) 45   e) 50   f) 55   g) 125   h) 280

8. Sum equals 5 times the middle number $n + (n + 1) + (n + 2) + (n + 3) + (n + 4) = 5n + 10 = 5(n + 2)$ where $(n + 2)$ is the middle number.

9. From 8 it follows at once that the sum of the five consecutive integers is an $(n + 2)$-multiple of 5.

10. a) $13 + \cdots + 17$ b) $17 + \cdots + 21$ c) $19 + \cdots + 23$ d) $33 + \cdots + 37$
e) $43 + \cdots + 47$ f) $49 + \cdots + 53$ g) $58 + \cdots + 62$ h) $68 + \cdots + 72$

11. None of them.

12. a) $3 + 4 + 5$
b) $7 + 8, 4 + 5 + 6, 1 + 2 + 3 + 4 + 5$
c) $5 + 6 + 7, 3 + 4 + 5 + 6$
d) $8 + 9 + 10, 13 + 14, 2 + 3 + \cdots + 7$
e) $9 + 10 + 11, 6 + 7 + 8 + 9, 4 + 5 + 6 + 7 + 8$
f) $17 + 18, 5 + 6 + 7 + 8 + 9, 2 + 3 + \cdots + 8$

g) $18 + 19$

h) $22 + 23$, $14 + 15 + 16$, $7 + 8 + \cdots + 11$, $5 + 6 + \cdots + 10$, $1 + 2 + \cdots + 9$

i) $17 + 18 + 19$, $12 + 13 + 14 + 15$, $2 + 3 + \cdots + 10$

j) $27 + 28$, $9 + 10 + \cdots + 13$, $1 + 2 + \cdots + 10$

k) $19 + 20 + 21$, $10 + 11 + \cdots + 14$, $4 + 5 + \cdots + 11$

l) $31 + 32$, $20 + 21 + 22$, $8 + 9 + \cdots + 13$, $6 + 7 + 8 + \cdots + 12$,
   $3 + 4 + \cdots + 11$

13. This problem is a rich investigation.

    If N is odd, the least number of consecutive integers needed is two.

    If N is even, the least number is described in a) and b) below.

    a) If N is even, get the prime factorization $N = 2^n p_1^{a_1} p_2^{a_2} \ldots p_s^{a_s} \ldots$

       Now if in the prime factorization, we have that $p_i > 2^{n+1}$ for all i (that is, each prime is greater than 2^{n+1}), then 2^{n+1} consecutive numbers are needed to sum to N. These consecutive numbers are found by taking $\frac{1}{2}$ of A where $A = p_1^{a_1} p_2^{a_2} \ldots, p_s^{a_s} + 1$ so that we have $\frac{A}{2}$.

       Now the term to the left of $\frac{A}{2}$ will be $(\frac{A}{2} - 1)$, such that $(\frac{A}{2} - 1) + \frac{A}{2}$ sum to $p_1^{a_1} p_2^{a_2} \ldots p_s^{a_s}$.

       Now take the number to the left of $(\frac{A}{2} - 1)$ and the one to the right of $\frac{A}{2}$ so that the sum equals $p_1^{a_1} p_2^{a_2} \ldots p_s^{a_s}$ and continue the process getting

       $$\cdots + \left(\frac{A}{2} - 2\right) + \left(\frac{A}{2} - 1\right) + \frac{A}{2} + \left(\frac{A}{2} + 1\right) + \cdots$$

    b) If N is even, get the prime factorization as in (a). If $p_i < 2^{n+1}$ for at least one i, there are p_i consecutive numbers needed for the sum N. The latter are determined by computing $\dfrac{N}{p_i}$ and listing the p_i consecutive numbers alternating above and below $\dfrac{N}{p_i}$.

EXERCISE 115

1. a) $10 + 11$ b) cannot be done   c) $11 + 12 + 13$ d) $22 + 23$
   e) $27 + 28$ f) $21 + 22 + 23$ g) $25 + 26 + 27$ h) $45 + 46$

2. The answers here are found by taking the sum of three consecutive integers $(1 + 2 + 3, 2 + 3 + 4, \ldots)$, then the sum of four consecutive integers, and so on.

   a) $0, 1, 2, 3, 4, 5, 6, 7, 8, 9$ b) $0, 2, 4, 6, 8$ c) $0, 5$
   d) $1, 3, 5, 7, 9$ e) $0, 1, 2, 3, 4, 5, 6, 7, 8, 9$
   f) $0, 2, 4, 6, 8$ g) $0, 1, 2, 3, 4, 5, 6, 7, 8, 9$ h) 5

EXERCISE 116

1. a) $10^2 = 9(11) + 1 = 100$ b) $11^2 = 10(12) + 1 = 121$
c) $14^2 = 13(15) + 1 = 196$ d) $21^2 = 20(22) + 1 = 441$
e) $26^2 = 25(27) + 1 = 676$ f) $31^2 = 30(32) + 1 = 961$

2. The pattern holds for any three consecutive integers n, $(n + 1)$, $(n + 2)$.
Square of middle integer $= (n + 1)^2 = n^2 + 2n + 1$.
The products of the other two integers plus $1 = n(n + 2) + 1 = n^2 + 2n + 1$.

3. $6 = 2 \times 3$, thus, for a number to be divisible by 6, there must be at least one factor of 2 and 3.
a) $20 = 2^2 \times 5$, $21 = 3 \times 7$ b) $26 = 2 \times 13$, $27 = 3^3$ c) $30 = 2 \times 3 \times 5$
d) $36 = 2^2 \times 3^2$ e) $42 = 2 \times 3 \times 7$
f) $56 = 2^3 \times 7$, $57 = 3 \times 19$

4. Yes, because in every set of three consecutive integers there is at least one multiple of 2 and exactly one multiple of 3. Since both factors of 6 are present, the product is divisible by 6.

EXERCISE 117

1. a) $360 + 1 = 361 = 19^2$ b) $840 + 1 = 841 = 29^2$
c) $1680 + 1 = 1681 = 41^2$ d) $3024 + 1 = 55^2$
e) $5040 + 1 = 5041 = 71^2$ f) $7920 + 1 = 7921 = 89^2$
g) $11880 + 1 = 11881 = 109^2$ h) $17160 + 1 = 17161 = 131^2$
i) $24024 + 1 = 24025 = 155^2$ j) $32760 + 1 = 32761 = 181^2$

2. Yes. $n(n + 1)(n + 2)(n + 3) = n^4 + 6n^3 + 11n^2 + 6n + 1 = (n^2 + 3n + 1)^2$, a perfect square.

3. a) $20 - 4 = 16 = 4^2$ b) $30 - 5 = 25 = 5^2$
c) $42 - 6 = 36 = 6^2$ d) $56 - 7 = 49 = 7^2$
e) $72 - 8 = 64 = 8^2$ f) $156 - 12 = 144 = 12^2$
g) $240 - 15 = 225 = 15^2$ h) $380 - 19 = 361 = 19^2$
i) $600 - 24 = 576 = 24^2$ j) $930 - 30 = 900 = 30^2$

4. Yes. $n(n + 1) - n = n^2 + n - n = n^2$, a perfect square.

EXERCISE 118

1. a) 123, 132, 213, 312; divisible by 1, 2, 3, 4
b) 246, 264, 426, 624; divisible by 1, 2, 3, 4
c) 135; divisible by 1
d) 036, 036, 603, 360, 630, 036, 063, 360, 063, 630; divisible by 1, 2, 3, ... , 10
e) 012, 012, 021, 120, 210, 120, 210, 120; divisible by 1, 2, 3, ... , 8
f) 369, 396, 639, 936; divisible by 1, 2, 3, 4

2.
1	1
1,2	2
1,2,3	$2(3) = 6$
1,2,3,4	$2^2(3) = 12$
$1, 2, \ldots, 5$	$2^2(3)(5) = 60$
$1, 2, \ldots, 6$	$2^2(3)(5) = 60$
$1, 2, \ldots, 7$	$2^2(3)(5)(7) = 420$
$1, 2, \ldots, 8$	$2^3(3)(5)(7) = 840$
$1, 2, \ldots, 9$	$2^3(3)^2(5)(7) = 2520$
$1, 2, \ldots, 10$	$2^3(3)^2(5)(7) = 2520$

A strategy: Begin with 1. 1 is divisible by 1 but not by 2. Thus, form the product 1(2). 2 is divisible by 1 and 2 but not by 3. Now multiply by 3 and get 1(2)(3) which is divisible by 1, 2, and 3 but not 4. Continuing $2^2(3) = 12$, and so on.

EXERCISE 119

1. It is possible to use the fact that the sum of n consecutive numbers beginning with $a \neq 1$ is given by $\dfrac{n(2a + n - 1)}{2}$. See also the answer to **EXERCISE 113**, problem 1.

a) $100 = 10^2$ b) $144 = 12^2$ c) $196 = 14^2$ d) $256 = 16^2$
e) $324 = 18^2$ f) $400 = 20^2$

2. The sum is the square of the number that is one more than the number of consecutive terms being added.

3. a) $81 = 9^2$ b) $121 = 11^2$ c) $169 = 13^2$ d) $225 = 15^2$
e) $289 = 17^2$ f) $361 = 19^2$

4. The sum is the square of the middle term and is also the square of the number of terms.

5. a) $5, 5^2$ b) $6, 6^2$ c) $7, 7^2$ d) $8, 8^2$ e) $9, 9^2$ e) $10, 10^2$

6. The sum is the square of the number of consecutive odd integers in the sum.

7. a) $16 - (1 + 3 + 5 + 7) = 0$, 4 odd numbers are needed with $\sqrt{16} = 4$ and $4^2 = 16$

b) $36 - (1 + 3 + \cdots + 11) = 0$, 6 odd numbers are needed with $\sqrt{36} = 6$ and $6^2 = 36$

c) $81 - (1 + 3 + \cdots + 17) = 0$, 9 odd numbers are needed with $\sqrt{81} = 9$ and $9^2 = 81$

d) $100 - (1 + 3 + \cdots + 19) = 0$, 10 odd numbers are needed with $\sqrt{100} = 10$ and $10^2 = 100$

e) $169 - (1 + 3 + \cdots + 25) = 0$, 13 odd numbers are needed with $\sqrt{169} = 13$ and $13^2 = 169$

8. To get the square root of a square number n^2, subtract the consecutive odd integers beginning with 1 until one gets 0. The number of subtractions is the square root of n^2.

EXERCISE 120

1. $5! = 120$ $6! = 720$ $7! = 5{,}040$ $8! = 40{,}320$
$9! = 362{,}880$ $10! = 3{,}628{,}800$ $11! = 39{,}916{,}800$

2. a) $4(5)(6) = 120$ b) 720 c) 56 d) 1 e) 20 f) 6 g) 20
    h) 126 i) 35

3. a) $(6!)\,(7!) = 10!$ implies $\dfrac{10!}{7!6!} = 1$ but $\dfrac{10!}{7!6!} = \dfrac{10 \times 9 \times 8}{6 \times 5 \times 4 \times 3 \times 2 \times 1} = 1$
    and the relation is valid.

    b) $(3!)(5!) = 6!$ implies $\dfrac{6!}{5!3!} = 1$ but $\dfrac{6!}{5!3!} = \dfrac{6}{3!} = 1$ and the relation is valid.

    c) $(3!)(5!)(7!) = 10!$ implies $\dfrac{10!}{3!5!7!} = 1$ but $\dfrac{10!}{3!5!7!} = \dfrac{(8)(9)(10)}{(6)(1)(2)(3)(4)(5)} = 1$
    and the relation is valid.

    d) $(4!)(23!) = (24)(23!) = 24!$

4. $(2!)(4!)(47!) = (2)(24)(47!) = (48)(47!) = 48!$

5.

1st	$\dfrac{2!}{2!(2-2)!} = 1$	6th $\dfrac{7!}{2!(7-2)!} = 21$
2nd	$\dfrac{3!}{2!(3-2)!} = 3$	7th $\dfrac{8!}{2!(8-2)!} = 28$
3rd	$\dfrac{4!}{2!(4-2)!} = 6$	8th $\dfrac{9!}{2!(9-2)!} = 36$
4th	$\dfrac{5!}{2!(5-2)!} = 10$	9th $\dfrac{10!}{2!(10-2)!} = 45$
5th	$\dfrac{6!}{2!(6-2)!} = 15$	10th $\dfrac{11!}{2!(11-2)!} = 55$

EXERCISE 121

1. There are $24 = (4!)$ ways.

    ABCD   *ABDC*   *ACBD*   *ACDB*   *ADBC*   *ADCB*   *BACD*   *BADC*
    BCAD   *BCDA*   *BDAC*   *BDCA*   *CABD*   *CADB*   *CBAD*   *CBDA*
    CDAB   *CDBA*   *DABC*   *DACB*   *DBAC*   *DBCA*   *DCAB*   *DCBA*

2. If the quilt is hanging, or has a trim which enables one to distinguish one corner from another, there are 16! ways to arrange the patches. However, without differentiating amongst the corners, there are only 16!/4 ways.

3. a) $5! + 2 = 122$ b) $5! + 3 = 123$
    c) $5! + 4 = 124$ d) $5! + 5 = 125$
    The answers are four consecutive composite numbers.

4. a) $6! + 2 = 722$ b) $6! + 3 = 723$ c) $6! + 4 = 724$
    d) $6! + 5 = 725$ e) $6! + 6 = 726$
    The answers are five consecutive composite numbers.

5. a) six   b) nine   c) 99

6. No. $n! + 2, n! + 3, \ldots, n! + n$ is a collection of $(n - 1)$ consecutive composite numbers. There is no limit to the length of the string of composite numbers; there is a problem writing the numbers in standard notation as the length of the string increases.

EXERCISE 122

1. 1 and 2 are single digit factorial sum numbers.

2. a) Yes   b) No

3. 0, 1, 2, 9 are the single digit subfactorial numbers.

4.

N	$N!$	$!N$	$!N + !N$
1	1	0	1
2	2	1	3
3	6	2	8
4	24	9	33
5	120	44	164
6	720	265	985
7	5040	1854	6894
8	40320	14833	55153

Answers will vary. For example, $N!$ is always even while $!N$ and $N! + !N$ are odd or even.

EXERCISE 123

1. All numbers from 3 through 14 are Ulam numbers.

3: 5 divisions	4: 2 divisions	5: 4 divisions	6: 6 divisions
7: 11 divisions	8: 3 divisions	9: 13 divisions	10: 5 divisions
11: 10 divisions	12: 7 divisions	13: 7 divisions	14: 12 divisions

It is quite probable that all positive integers are Ulam numbers. However, what is important to investigate is the *number* of divisions required for each integer to reach 1.

2.

Number	Steps to reach 1	Peak	Number	Steps to reach 1	Peak
1	3	4	14	17	52
2	1	2	15	17	160
3	7	16	16	4	8
4	2	4	17	12	52

Number	Steps to reach 1	Peak	Number	Steps to reach 1	Peak
5	4	16	18	20	52
6	8	16	19	20	58
7	16	52	20	7	16
8	3	4	21	7	64
9	19	52	22	15	52
10	6	16	23	15	160
11	14	52	24	10	16
12	9	16	25	23	88
13	9	40			

Note the repetitions of peak values.

EXERCISE 124

1. Several choices are possible.

a) 33; 44; 55; 66

b) 121; 232; 343; 454

c) 1221; 4554; 1331; 2222

d) 12,321; 56, 765; 25,352; 69,796

e) 256,652; 897,798; 543,345; 666,666

2. a) 90   b) 90   c) 900

3. Several choices are possible. One possibility is provided.

a) 2, 22, 222, 2222, 22222, 222222, 2222222, 22222222, 222222222, 2222222222

b) 4, 44, 464, 4664, 46464, 464464, 4646464, 46466464, 464646464, 4646446464

c) 7, 77, 707, 7007, 70707, 707707, 7070707, 70700707, 707070707, 7070770707

4. a) $5 = 1 + 3 + 1 = 1 + 1 + 1 + 1 + 1$

b) $6 = 3 + 3$ $\qquad = 1 + 4 + 1 \qquad = 2 + 2 + 2$

$\quad = 1 + 2 + 2 + 1 \qquad = 2 + 1 + 1 + 2 = 1 + 1 + 2 + 1 + 1$

$\quad = 1 + 1 + 1 + 1 + 1 + 1$

c) $7 = 1 + 5 + 1 \qquad = 2 + 3 + 2 = 3 + 1 + 3$

$\quad = 1 + 1 + 3 + 1 + 1$

$\quad = 1 + 1 + 1 + 1 + 1 + 1 + 1$

d) $8 = 4 + 4$ $\qquad = 1 + 6 + 1 \qquad = 2 + 4 + 2$

$\quad = 3 + 2 + 3 \qquad = 1 + 3 + 3 + 1 \qquad = 2 + 2 + 2 + 2$

$\quad = 3 + 1 + 1 + 3 \qquad = 1 + 1 + 4 + 1 + 1 = 2 + 1 + 1 + 1 + 1 + 2$

$\quad = 1 + 2 + 1 + 1 + 2 + 1 = 1 + 1 + 2 + 2 + 1 + 1$

$\quad = 1 + 1 + 1 + 2 + 1 + 1 + 1$

$\quad = 1 + 1 + 1 + 1 + 1 + 1 + 1 + 1$

EXERCISE 125

1. a) 55,   1 reversal   b) 121,   2 reversals
   c) 121,   2 reversals   d) 121,   2 reversals
   e) 1111,   3 reversals   f) 4884,   4 reversals
   g) 1111,   3 reversals   h) 4884,   4 reversals
   i) 44044,   6 reversals   j) 1111,   3 reversals
   k) 4884,   4 reversals   l) 44044,   6 reversals
   m) 8813200023188,   24 reversals   n) 88555588,   11 reversals
   o) 233332,   8 reversals   p) 233332,   8 reversals
   q) 233332,   8 reversals   r) 99099,   5 reversals
   s) 99099,   5 reversals   t) 67276,   4 reversals

2. Several answers are possible.   a) 3.3, 4.4, 5.5, 6.6
   b) 45.54, 13.31, 22.22, 15.51   c) 256.652, 897.798, 543.345, 666.666

3. a) 3.3,   1 reversal   b) 13.31,   2 reversals   c) 15.51,   2 reversals
   d) 1887.7881,   7 reversals   e) 39.93,   2 reversals   f) 11.11,   2 reversals
   g) 1476.6741,   7 reversals   h) 35.53,   2 reversals   i) 329.923,   5 reversals
   j) 59.95,   1 reversal

4. Of the 90 two-digit numbers 10–99, 53 require 1 reversal, 22 require 2 reversals, 5 require 3 reversals, 4 require 4 reversals, 4 require 6 reversals, and 2, namely 89 and 98, require 24 reversals to reach a palindrome.

EXERCISE 126

1. Lulu, Lee, Faith, Molly, Aika, Guri are palindromic names.
2. Results will vary.
3. December
4. None
5. 2222, 2332, 2442, 2552, 2662
6.

```
       111          11111111           11111
      11211        1122222211          1122211
     1121211      112233332211        112232211
    112121211    11223344332211      11223232211
   11212121211                      1122323232211
      (a)            (b)                (c)
```

EXERCISE 127

1. 66, 171, 595, 666, 3003, 5995, 8778. A fairly large triangular number that is also palindromic is 61728399382716.
2. $22^2 = 484$, $101^2 = 10201$, $111^2 = 12321$.
3. $101^3 = 1030301$, $111^3 = 1367631$, $1001^3 = 1003003001$.

4. a) 698896, yes   b) 1048576, no   c) 104060401, yes
    d) 637832238736, yes   e) 4461012876321, no   f) 4099923883299904, yes

EXERCISE 128

1. 191, 313, 353, 373, 383

2. a) 13 and 31,   17 and 71,
    37 and 73   79 and 97

   b) 113 and 311,   107 and 701,   149 and 941,
     157 and 751,   167 and 761,   179 and 971

   c) 1033 and 3301,   1009 and 9001,   1021 and 1201,
     1031 and 1301,   1061 and 1601,   1069 and 9601

   d) 10061 and 16001,   10067 and 76001,   10069 and 96001,
     10079 and 97001,   10091 and 19001,   10151 and 15101

3. 199, 919 and 991,   337, 733 and 373

4.
1009	1021	1031	1033	1061	1069	1091	1097	1103
1109	1151	1153	1181	1193	1201	1213	1217	1223
1229	1231	1237	1249	1259	1279	1283	1301	1321
1381	1399	1409	1429	1439	1453	1471	1487	1499
1511	1523	1559	1583	1597	1601	1619	1657	1669
1723	1733	1741	1753	1789	1811	1831	1847	1867
1879	1901	1913	1933	1949	1979			

EXERCISE 129

1. $B_5 = 52$ $B_6 = 203$ $B_7 = 877$ $B_8 = 4,140$
   $B_9 = 21,147$ $B_{10} = 115,9575$ $B_{11} = 678,570$ $B_{12} = 4,213,597$

2. Given: 4 lines of poetry. Same letter in a column means these lines rhyme.

Line 1	*a*	*a*	*a*	*a*	*b*	*a*	*a*	*a*	*a*	*a*	*a*	*a*	*a*	*a*
Line 2	*a*	*a*	*a*	*b*	*a*	*a*	*a*	*b*	*b*	*b*	*b*	*b*	*b*	*b*
Line 3	*a*	*a*	*b*	*a*	*a*	*b*	*b*	*a*	*a*	*c*	*b*	*b*	*c*	*c*	*c*
Line 4	*a*	*b*	*a*	*a*	*a*	*c*	*b*	*c*	*b*	*a*	*a*	*c*	*b*	*c*	*d*

3. Let the 4 customers be *A B C D*. Let the cars be marked car 1, car 2, car 3, car 4. Then we have 15 arrangements displayed.

Car 1	*ABCD*	*ABC*	*ABD*	*ACD*	*BCD*	*AB*	*AB*	*AC*	*AC*
Car 2	—	*D*	*C*	*B*	*A*	*C*	*CD*	*B*	*BD*
Car 3	—	—	—	—	—	*D*	—	*D*	—
Car 4	—	—	—	—	—	—	—	—	—

Car 1	AD	AD	A	A	A	A
Car 2	B	BC	BC	BD	B	B
Car 3	C	—	D	C	CD	C
Car 4	—	—	—	—	—	D

4. Let the donuts be J, C, H and the plates be P_1, P_2, and P_3. There are 5 arrangements.

P_1	JCH	JC	JH	CH	J
P_2	H	C	J		C
P_3					H

Let the donuts be J, C, H , O and the plates be P_1, P_2, P_3, and P_4. There are 15 possibilities as enumerated in 2 and 3.

EXERCISE 130

1. The total number of ways $C_4 = 14$.

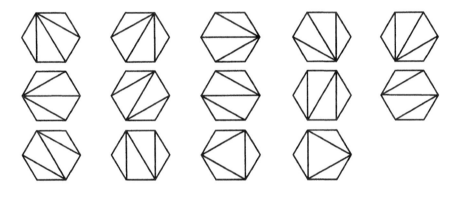

2.

$$\begin{array}{cccc} 1 & 1 & 2 & 5 \\ \times 5 & \times 2 & \times 1 & \times 1 \\ \hline 5 & +2 & +2 & +5 = 14 = C_4 \end{array}$$

$$\begin{array}{ccccc} 1 & 1 & 2 & 5 & 14 \\ \times 14 & \times 5 & \times 2 & \times 1 & \times 1 \\ \hline 14 & +5 & +4 & +5 & +14 = 42 = C_5 \end{array}$$

$C_6 = 132 \quad C_7 = 429 \quad C_8 = 1,430 \quad C_9 = 4,862 \quad C_{10} = 16,796$
$C_{11} = 58,786$

3. Six people:

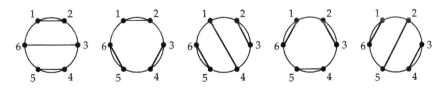

Eight people:

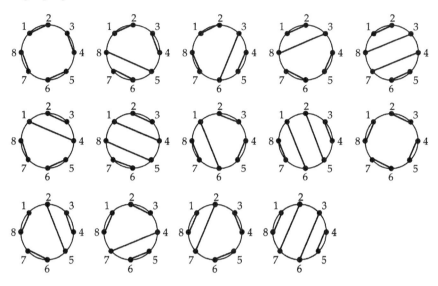

4. Let the dancers from shortest to tallest be $1, 2, 3, 4, 5, 6$. There are 5 arrangements.

456   356   346   246   256
123   124   125   135   134

Let the dancers from shortest to tallest be 1, 2, 3, 4, 5, 6, 7, 8. There are 14 arrangements.

5678   4678   4578   4568   3678
1234   1235   1236   1237   1245

3578   3568   3478   3468   2678
1246   1247   1256   1257   1345

2578   2568   2478   2468
1346   1347   1356   1357

There is a formula for Catalan numbers: $C_n = \dfrac{(2n)!}{n!(n+1)!}$.

EXERCISE 131

1. $F_{11} = 89,$ $F_{12} = 144,$ $F_{13} = 233,$ $F_{14} = 377,$ $F_{15} = 610,$
   $F_{16} = 987,$ $F_{17} = 1597,$ $F_{18} = 2584,$ $F_{19} = 4181,$ $F_{20} = 6765$

2. The number of Fibonacci numbers is unlimited. This follows from the definition and rule of formation above. We can always add the preceding two numbers to get a third Fibonacci number.

3. Sequence: 5, 6, 11, 17, 28, 45, 73, 118, 191, 309. 7th term = 73.

  Sum = 803 = 11 × 73.

4. Let the first number be a, the second b.

Term	Fibonacci-like Sequence
1	a
2	b
3	$a + b$
4	$a + 2b$
5	$2a + 3b$
6	$3a + 5b$
7	$5a + 8b$
8	$8a + 13b$
9	$13a + 21b$
10	$21a + 34b$

Sum = $55a + 88b = 11 \times (5a + 8b)$

5.

5	12	1	$F_7 = 13$
6	20	1	$F_8 = 21$
7	33	1	$F_9 = 34$
8	54	1	$F_{10} = 55$
9	88	1	$F_{11} = 89$
10	143	1	$F_{12} = 144$

6. The sum of the first n Fibonacci numbers is $F_{n+2} - 1$.

The Fibonacci Quarterly, the official journal of the Fibonacci Association, began publishing in February 1963 with Volume 1, Number 1. The quarterly is a rich source of information on Fibonacci numbers, Lucas numbers and related topics.

EXERCISE 132

1.

5	26	3	$L_7 = 29$
6	44	3	$L_8 = 47$
7	73	3	$L_9 = 76$
8	120	3	$L_{10} = 123$
9	196	3	$L_{11} = 199$

2. The sum of the first n Lucas numbers is $L_{n+2} - 3$.

EXERCISE 133

1. $T_{11} = 274,$ $T_{12} = 504,$ $T_{13} = 927,$ $T_{14} = 1,705,$
   $T_{15} = 3,136,$ $T_{16} = 5,768,$ $T_{17} = 10,609,$ $T_{18} = 19,513,$
   $T_{19} = 35,890,$ $T_{20} = 66,012$

2. a) 3   b) 4   c) 5   d) 6   e) 6   f) 8   g) 14   h) 9
   i) 16   j) 10   k) 11   l) 17   m) 12   n) 13   o) 15

3. All differences are 0 on row 6.

4. All differences are 0 on row 6.

5. This is a rich investigation. The application of the tribonacci numbers dis-
cussed here can be found in S. Bezuszka, and L. D'Angelo. "An Application
of Tribonacci Numbers." *Fibonacci Quarterly*, Vol. 15, No. 2 (April 1977),
pp. 140–144. See also *Fibonacci Quarterly*, Vol. 1, No. 3 (October 1963), pp.
71–74, where the term tribonacci (perhaps also the term tetranacci) seems to
have been coined by Mark Feinberg when he was fourteen years old.

EXERCISE 134

1. $Q_{11} = 401,$ $Q_{12} = 773,$ $Q_{13} = 1,490,$ $Q_{14} = 2,872,$
   $Q_{15} = 5,536,$ $Q_{16} = 10,671,$ $Q_{17} = 20,569$ $Q_{18} = 39,648$
   $Q_{19} = 76,424,$ $Q_{20} = 147,312$

2. *b, d, e, f*

3. The first ten terms in the phibonacci number sequence are: 1, 2, 3, 5, 7, 11, 17,
23, 37, 41.

For more information on phibonacci numbers, see A. Bager, "Problem E2833."
American Mathematical Monthly, Vol. 87: 5 (May 1980), p. 404.

EXERCISE 135

1. *a* and *c* are survivor numbers.

2. Answers will vary. Some possibilities:
   a) $13 + 69 = 82$ b) $82 - 69 = 13$ c) $11 \times 57 = 627$ d) $627/11 = 57$

EXERCISE 136

1. a) 3838   b) 4646   c) 9999   d) 134134
   e) 240240   f) 999999   g) 251251   h) 365365

2. $abc \times 7 \times 11 \times 13 = abc \times 1001 = abcabc$

3.

	÷7 =	÷11 =	÷13 =
a)	33891	3081	237
b)	65208	5928	456
c)	98527	8957	689
d)	126984	11544	888
e)	135278	12298	946

4. Yes, because $abcabc = abc\ (7 \times 11 \times 13)$ so $\dfrac{abcabc}{7 \times 11 \times 13} = abc$

EXERCISE 137

1. a) 10(13)   b) 7(22)   c) 4(55) = 10(22)
   d) 4(58)   e) 4(94)   f) 4(100) = 10(40) = 16(25)
   g) 4(115) = 10(46)   h) 4(121) = 22(22)

2. The set of Lagado numbers is generated by the formula $(3n - 2)$.

Index

Integers are the fountainhead of all mathematics.

Hermann Minkowski (1859–1930)

Families of numbers are listed individually; "number" is suppressed from the listings.

Printed in the United States
By Bookmasters

Printed in the United States
By Bookmasters